WATER AND WIND POWER

Martin Watts

Shire Publications

Published in 2000 by Shire Publications Ltd, Cromwell House, Church Street, Princes Risborough, Buckinghamshire HP27 9AA, UK. Website: www.shirebooks.co.uk

British Library Cataloguing in Publication Data: Watts, Martin. Water and wind power. 1. Wind power – Great Britain – History 2. Wind power – Ireland – History 3. Water-power – Great Britain – History 4. Water-power – Ireland – History I. Title 333.9'14. ISBN 0 7478 0418 4.

Front cover: (Top) The smock mill at Sarre, Kent. (Bottom) The watermill at Rossett, near Wrexham, North Wales.
Back cover: (Top) Waterwheel at Newnham Mill, Newnham Bridge, Worcestershire. (Bottom) The tower mill at Thelnetham, Suffolk.

Printed in Great Britain by CIT Printing Services, Press Buildings, Merlins Bridge, Haverfordwest, Pembrokeshire SA61 1XF.

CONTENTS

The post mill at Aythorpe Roding, Essex, now in the care of Essex County Council and maintained in full working order.

ACKNOWLEDGEMENTS

In the thirty years that I have been interested in and involved with mills, I have met many people who have almost without exception been willing to share information and allow me access to their properties. I would like to acknowledge their generosity and also convey my thanks to many friends and colleagues, past and present, including those in the Mills Section of the Society for the Protection of Ancient Buildings, the Association for Industrial Archaeology and the Traditional Corn Millers Guild, all of whom have helped broaden my understanding and appreciation of mills. Particular thanks are due to Tim Booth, Stephen Buckland, Mildred Cookson, David Crossley, Joan Day, Peter Dolman, Roy Gregory, John Harrison, John Langdon, Michael Lewis, Ken Major, Vincent Pargeter, Colin Rynne, Peter Stanier and James Waterfield for help with information, references and illustrations, and I am grateful to English Heritage and the National Trust for allowing me to use survey material commissioned by them. I must also express my gratitude to Shire Publications for asking me to write this in the first place and for their forbearance whilst I took the time to do so. That it has been written at all is in no small measure due to the unstinting support of Alan Stoyel, who has generously shared his mill archive and whose critical perception I have always found valuable and instructive, and Sue, my wife, who has helped with research and fieldwork and is my most constructive critic.

Martin Watts 1999

Illustrations are taken from material in the author's own collection, with a number of exceptions, which are acknowledged as follows: David Crossley, page 35; Peter Dolman, page 82 (bottom); David H. Jones, page 96; The Royal Society, page 71; Colin Rynne, page 12; SPAB Mills Section, pages 54 (bottom), 83 (bottom), 104; Alan Stoyel, pages 50 (top), 52, 59, 73, 74 (bottom), 79 (top), 120; Derrick Warren, page 113.

INTRODUCTION

The study of mills, particularly those driven by natural sources of power, has developed considerably since the word 'molinology' was first used in 1962. This general introduction sets out to summarise the development and use of water and wind power in the British Isles from Roman times to the present, using some of the wealth of information that has been made available by archaeology, historical research and fieldwork.

Throughout the book watermills and windmills are considered as parts of the same development, both technically and historically, for it seems misleading to separate one from the other, particularly as machinery, whether powered by animals, water or wind, was usually built and run by the same people. It is also dangerous to separate windmills from watermills for preservation purposes; both are equally vulnerable to destruction by natural elements, as well as by reuse and other development. Windmills have often been considered more significant than watermills, partly because they are usually visually exciting and in prominent locations, but the shaping of the landscape and the development of industry in the British Isles have been influenced much more by the use of water power than of wind power.

Mills should also be seen in context as an essential part of the historical fabric of the built environment. They were engines made to perform useful

The power of water: a weir on the river Teign, Devon, with a piped leat taken off on the left-hand side.

work, initially to process foodstuffs and supply water, later to work metals, produce cloth and serve a great number of other trades and industries that were vital to the social and economic development of Britain in the eighteenth and nineteenth centuries. Even steam engines, which came to supplant the use of natural power in the nineteenth century, could not have been built in the first place without waterwheels to drive the machinery that was needed to smelt and work the iron and turn cylinders. This account is not concerned merely with the technology of using water and wind, but also with some of the personalities who were involved with that development and the reasons that lay behind it.

Historically there is a great richness and variety of water- and wind-powered usage and, although much has been lost, sufficient examples remain to provide a useful picture of this important aspect of Britain's technological development. There are still new discoveries to be made and a great deal of fieldwork and recording to be done. Many small water- and wind-powered corn mills survive, some complete and in working order, but the large textile mills of the eighteenth and nineteenth centuries are perhaps the most imposing remains and their protection as historic buildings is just one of the problems that face conservationists. There is also ample evidence of the mills and machinery used in the extractive and metal-processing industries in many upland regions, including the National Parks, which now provide important recreational areas. The development of water and wind power has not ended and modern uses such as electricity generation indicate the importance of natural, renewable power sources whose value to mankind has been recognised for over two thousand years.

Many of the technical terms used in this book are explained in the Glossary, beginning on page 122. For the guidance of readers, the first time each such term is used in the text it is printed in italic type.

THE INTRODUCTION OF WATER POWER INTO BRITAIN

Roman watermills

The origins of water power and the use of waterwheels to transform the energy of flowing or falling water to do useful work have long been debated and will undoubtedly continue to be so as more evidence is found. Waterwheels were probably first used in south-east Europe and the Near East by the second century BC for raising water for irrigation and, shortly after, for driving stones for milling grain. The Roman architect Vitruvius, in the tenth book of his treatise on architecture, written probably between 25 and 23 BC, describes a number of machines, including treadwheels for hoisting, waterwheels for raising water and a watermill. While the argument over his description of the gears that drove the *millstones* cannot easily be resolved, that he describes a waterwheel rotating in a vertical plane on a horizontal axis is indisputable. This form of wheel, generally termed the *vertical waterwheel,* was introduced into Britain by the Romans during the first century AD.

The first Roman watermill site to be identified in Britain was found at Haltwhistle Burn Head, close to Hadrian's Wall in Northumberland, in 1907. Archaeological excavation revealed a stone-built mill with a timber-lined wheel race. Although the site is now totally obscured, the substantial millstone fragments that were found are on display in the museum at Chesters Roman Fort, Northumberland, along with many examples of *querns,* the smaller hand-driven stones which were the usual means of grinding grain before the introduction of watermills. Other possible sites have been identified close to Hadrian's Wall, connected with the bridge abutments at Chesters Fort, on the North Tyne, and further west at Willowford, Cumbria, on the river Irthing, both places where the wall crossed major rivers. At neither location has positive evidence of a mill been found, but watercourses and sluice works suggest strongly that waterwheels, perhaps used for milling grain, were sited close by. The Haltwhistle Burn Head mill has been dated to the third century AD, but its location and the shape of the millstones found there suggest that it may be of earlier origin. The earliest Roman watermill so far discovered in Britain is that at Ickham, Kent, where an *earth-fast* timber-framed structure was found close to the Little Stour river during gravel extraction in the 1970s. It was probably working by AD 150 and was in use for about a hundred and thirty years. A little further upstream, to the west, a second watermill site, of fourth-century date, was found.

Although the number of positively identified Roman watermills in Britain is few, it is likely that power-driven mills were more common than once thought. Evidence of milling and baking has been found at many sites, and large granaries, both military and civilian, indicate the importance of corn crops and their processing to the people of Roman Britain and, in particular, to the

Bridge abutment at Willowford, Cumbria, where Hadrian's Wall crossed the river Irthing. The sluice ways may have been connected with a Roman watermill site.

army. There is also evidence that grain was milled at urban bakeries, in *Londinium* (London) and *Calleva Atrebatum* (Silchester), for example, although nothing on the scale of the bakeries at Pompeii, with their donkey-powered 'hourglass' mills, has yet been discovered. Some fragments of mills of this type have been found in Britain. There was one in London that has been reconstructed as part of a display on Roman milling at the Museum of London and is exhibited along with two finely made lava millstones found in the vicinity of the river Walbrook, the stream that flowed through the Roman city. These are dated to the first or second century AD and are probably from one or more watermills.

Archaeological excavation of a large villa, one of the earliest to be established in Britain, at Gadebridge Park, Hemel Hempstead, Hertfordshire, revealed a *leat,* dating apparently to the late first century AD, which had been redirected early in the fourth century. It may have served a watermill as well as baths, for here, as at some other sites, fragments of millstones were found. In many instances such fragments have been identified as being from querns, but some are from stones too large in diameter to have been hand-turned, at least without some form of intermediate gearing. Querns are common finds on all types of Roman sites, and many examples are on display in museums, sometimes mislabelled, for the distinction between querns and millstones has always caused problems. There is no hard and fast rule, but available evidence suggests that stones with a diameter of over 2 feet (0.6 metre) are more likely to be from power-driven mills.

At Woolaston, Gloucestershire, two millstones that had been reused in an early fourth-century floor were discovered when a villa site was excavated in the 1930s. One is now in the collection of Gloucester City Museum and Art Gallery and shows features in common with other Roman millstones. Its size, shape and the double dovetail-shaped slots by which it was located and rotated indicate that it was power-driven. Similarly at Chedworth, an extensive villa site also in Gloucestershire, a millstone on display is too large for hand operation. Moreover, the interesting pattern of *dressing* on the working face appears to reflect an understanding of the function of the different parts of the grinding surface of the stone when in use. It is coarsely pecked close to the *eye* (the hole through its centre), to break open the grain, then cut with a series of almost radial furrows from there to its periphery, to reduce the broken grain into meal. Roman millstones are generally well dressed, often with a pattern of furrows laid out in triangular segments similar to those found on millstones in recent use and known as *harps*. The function of dressing is to provide a series of cutting edges, to allow the stones to perform more work and to give more control when milling. The way millstones are dressed has an effect on both output and product: the Romans used both wholemeal and sieved wheat flour. The benefits of wholemeal to the digestive system were apparently well known even then.

Little physical evidence has been found of the waterwheels built and used in Roman Britain, but they were predominantly of timber, with iron fastenings and *journals* that turned in stone bearings. They seem generally to have been *undershot* and to have been located in timber-lined wheelpits or *races*, fed by man-made channels or leats. From sites where information is available, they

Roman millstone, Chedworth, Gloucestershire. The milling face is divided into two zones, to break open the grain near the centre and to reduce it by the cutting action of the furrows.

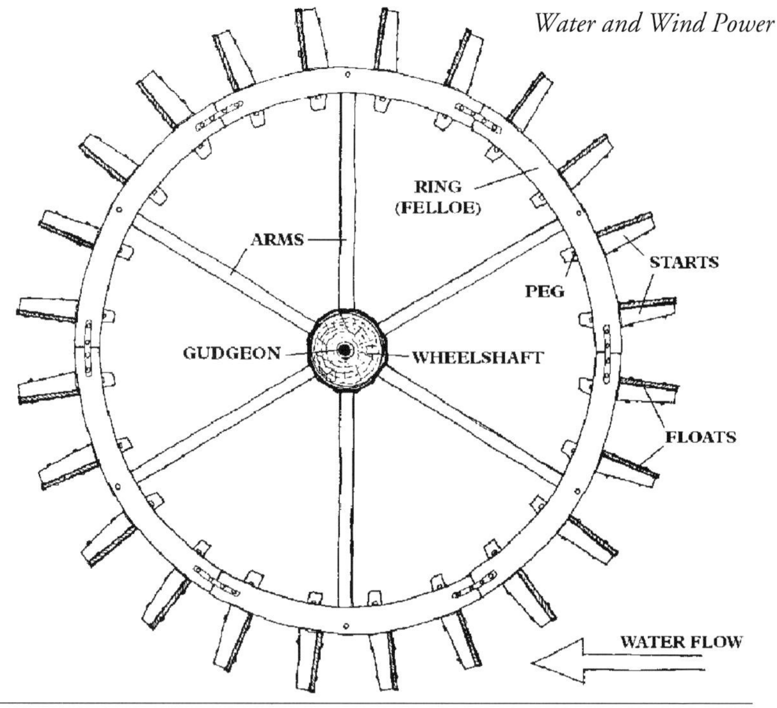

Simple undershot waterwheel, with naming of parts.

were from 6 to 11 feet (2.0 to 3.5 metres) in diameter by about 1 to 2 feet (0.3 to 0.6 metre) wide. Power output was low, probably no more than 1 horsepower (0.75 kW), but the millstones associated with these wheels are generally of relatively small diameter and the output of mills has always been dependent on a number of highly variable factors. No waterwheel components have yet been discovered, although two wrought-iron millstone *spindles* have been identified, one in a hoard of metalwork discovered down a well at Great Chesterford, Essex, the other from *Calleva Atrebatum* (Silchester). It is not known, however, if these came from watermills or some form of geared hand- or animal-driven mill. Millstones have been found in a number of non-working contexts, such as hearths and floors, and by the fourth century AD several known Roman mill buildings were deserted or reused. A notable instance of the latter was excavated in 1990 at the Redlands Farm Villa, Stanwick, Northamptonshire, on the flood plain of the river Nene, where the mill, which was part of a barn, became the core of a winged corridor villa – an early example of a mill conversion.

The majority of Roman water-power sites that have been identified were used for grain milling, but there is some evidence to suggest that waterwheels were also used for industrial purposes. The fourth-century site at Ickham, Kent, probably had more than one waterwheel, and in an area adjacent to the corn mill great quantities of metal and numerous artefacts of iron, pewter and copper alloys were discovered. Among the finds were *bearing* stones and an iron hammer head that showed mechanical wear, which suggests that water power may have been used for metalworking there. Similarly there is a possibility that water power was used to drive ore-crushing hammers or stamps at the Roman gold mine at Dolaucothi in south-west Wales by the end of the first century AD, and further evidence undoubtedly awaits discovery elsewhere.

The Roman achievement in building and civil engineering has long been acknowledged, and much evidence of hydraulic works in terms of leats and aqueducts can still be found. It is likely that further water-power sites will be identified by archaeologists, particularly when exploring the areas surrounding forts, settlements and villas, many of which were located close to substantial natural watercourses. Water has always played a major role in the siting of human settlement and it is inconceivable that a people as well organised and as technologically competent as the Romans would not have made full use of its power in Britain.

Anglo-Saxon and early Irish watermills

The first historical reference to a watermill in Britain occurs in an Anglo-Saxon charter dated to AD 762, between Aethelberht, king of Kent, and the minster of Saint Peter and Saint Paul at Canterbury, whereby the community gave half-use of a mill at Chart, Kent, to the royal vill (manor) at Wye, in return for pasture rights in the Weald for their tenant at Chart. The same, or another, mill at Chart is recorded in 814, and from the ninth century on there is an increasing number of references to mills in charters and place-names. The term 'millwheel' first occurs in the tenth century, while the more familiar 'waterwheel' is not used until later in the Middle Ages. Similarly, there are several different Anglo-Saxon terms for weirs, dams and watercourses, suggesting that dams were built both to impound water in ponds and to divert water into artificial courses from natural ones. It is likely that tidal power was also used, for a charter dated 949 appears to refer to a *tide mill* near Reculver, Kent. Certainly by the eleventh century watermilling was well established in England, for the Domesday Survey, compiled for William I in 1086 and reflecting the state of the country at the time of the Norman Conquest, records over six thousand mills.

The archaeological evidence for mills in the Anglo-Saxon period is slim and often conflicting. Significant remains of both vertical- and *horizontal-wheeled* mills of early-medieval date have been found in Ireland, however. Of nearly one hundred horizontal-wheeled mills so far identified, about half have been dated by *dendrochronology* from the seventh to the eleventh centuries. Eight parts of a mill were named in an Irish legal tract written before the end of the sixth century, and by the seventh century there was an Irish word for a *millwright*, then considered to be a craftsman equivalent in status to the lowest grade of nobility, in contrast to the lowly position of the slave-women the kings of Kent employed to work at the quern for them at the same period in history. The introduction of the watermill in Ireland was said by an eleventh-century Irish poet to have been at the instigation of King Cormac in the third century, to release a beautiful bondmaid from the impossible task of grinding by hand nine quarters of corn a day, a task set her by Cormac's queen both to weary her and to keep her safely employed in the bakery, out of the king's way. Cormac is said to have sent across the sea (in which direction is not recorded) to bring mechanics who could build a watermill, but there is no historical or archaeological proof for this delightful story.

The early-medieval Irish watermills were framed of massive oak timbers laid on the ground, with one or sometimes two horizontal waterwheels located beneath the building. These wheels averaged about 3 feet (1 metre) in diameter and usually had nineteen paddles, each wheel driving a single pair of millstones from 1 foot 9 inches to 3 feet (0.5 to 1 metre) in diameter. The waterwheel and upper millstone were directly connected by a vertical shaft or spindle, so that the stone ran at the same speed as the wheel. The bearing at the foot of the spindle was of stone set into a timber beam, and water was directed on to the paddles of the wheel through an inclined timber *penstock*. Water usually entered the penstock from a millpond that was formed by

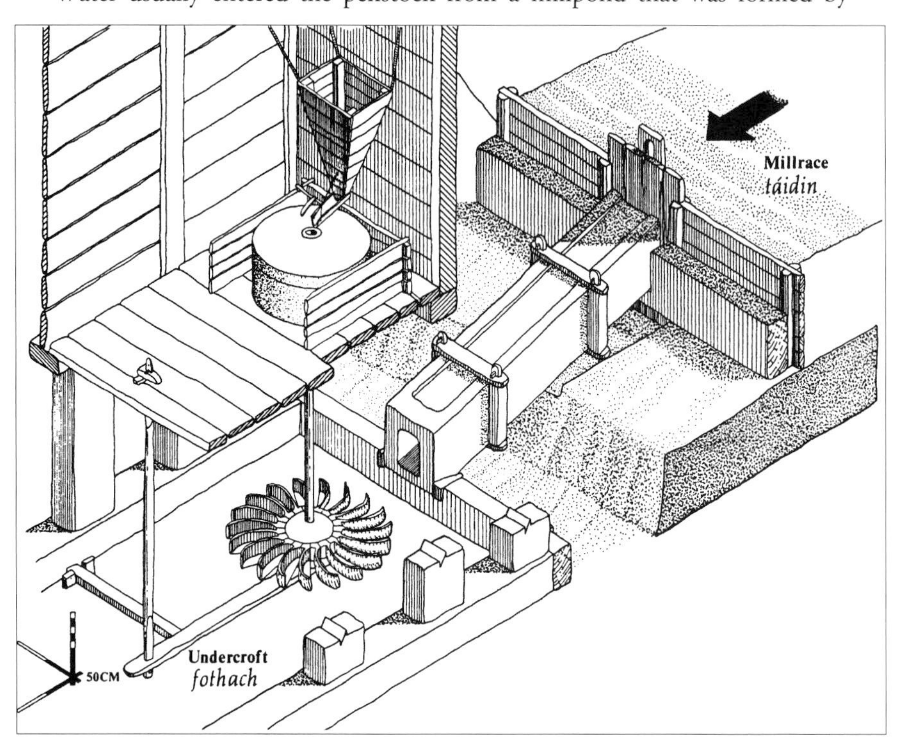

Reconstruction drawing by Colin Rynne of the ninth-century horizontal-wheeled watermill at Cloontycarthy, County Cork.

building an earth dam, often revetted with stones and perhaps timber boards, across a stream, but the earliest mill site found in Ireland appears to have exploited the rise and fall of the tide to give a working head of water. At Little Island, County Cork, timbers that have been dendrochronologically dated to about 630 were excavated in 1979 and interpreted as being from a horizontal-wheeled mill, apparently with two wheels, with a vertical-wheeled mill close by. The mill structures were built on oak piles, and archaeological evidence showed that the action of the tide, as well as powering the waterwheels, had eventually accounted for their demise. Another vertical-wheeled mill site has been found at Morett, County Laois, where a timber-lined race for a wheel up to 13 feet (4 metres) in diameter by 2 feet (0.6 metre) wide has been dated to the late eighth century. From present evidence it appears that the majority of the horizontal-wheeled mills discovered in Ireland were built in a remarkably short period, between 770 and 880.

In England a significant find was made in 1971 when the remains of a Saxon watermill, possibly attached to the Mercian royal palace at Tamworth, Staffordshire, were excavated. Two generations of horizontal-wheeled mill were found to have existed virtually on the same site, and timbers have been dated to the middle of the ninth century, which coincides with Tamworth's most important period as a Saxon town and is also within the main period of watermill-building in Ireland. The mill was supplied with water from a partially timber-lined pond, fed by a leat from the river Anker. Finds included a paddle from the waterwheel, a steel bearing that was worn on both sides, and numerous millstone fragments, some of German lava but mostly of local sandstone. The flat grinding faces of the millstones were not dressed with furrows in the Roman manner but showed signs of having been pecked, to produce numerous cutting edges. Similarly dressed millstone fragments have been found at other possible Saxon mill sites, including Wharram Percy, North Yorkshire, where a Saxon dam and pond that undoubtedly served a watermill have been excavated.

The other important early watermill site found in England was at Kingsbury, Old Windsor, Berkshire, in 1957. Here a great leat some 1¼ miles (2 km) long and nearly 20 feet (6 metres) wide had been dug across the neck of a loop in the Thames, to drive three vertical waterwheels working in parallel. The timbers of the mill, which lay below the present water table and were well preserved, have been dated to the late seventh century. After this impressive mill was destroyed in the ninth or tenth century, the leat was filled and a narrow channel cut into it, to supply water to a horizontal-wheeled mill. The smaller leat was apparently recut several times before going out of use early in the eleventh century. Like that at Tamworth, this mill may have been attached to a royal palace.

The existence of both vertical- and horizontal-wheeled mills alongside each other in the early Middle Ages is an interesting problem and the argument that one form preceded the other is difficult to resolve. That the Romans used the vertical wheels with gearing to turn the drive from a vertical plane to a horizontal one in order to drive millstones is now beyond question. While it is likely that the vertical-wheeled geared mill survived after the Romans left Britain in the early fifth century, its mechanical complexity and relatively high maintenance requirement would probably have worked against its continued use during the Anglo-Saxon period. It has been observed that the mechanically simple horizontal-wheeled mill is a product of peasant culture and that its construction and running costs were low. The remains found at Tamworth and Old Windsor are important evidence of the existence of horizontal-wheeled mills in England. They should be considered along with a chance find near Nailsworth, Gloucestershire, of some paddles from a horizontal waterwheel, which have not been dated, and the identification of some timbers on the bank of the river Tyne, near Corbridge, Northumberland, which are thought to be part of the substructure of a horizontal-wheeled mill.

It has been suggested that the lower-value mills recorded in Domesday Book had horizontal waterwheels, although as yet there is neither documentary nor archaeological evidence to support this. It is perhaps not unreasonable to suggest that in England watermills with horizontal wheels were being systematically replaced by both vertical-wheeled mills and windmills by the end of the twelfth century. The earliest depictions of watermills and the increased documentary evidence of expenditure on mills, which both occur during the thirteenth century, show that within two hundred years of the Norman Conquest English manorial mills had vertical waterwheels.

Watermills with horizontal wheels

Horizontal-wheeled mills are sometimes referred to as Norse mills because they are found in areas once settled by people from Scandinavia. As well as the

Horizontal-wheeled watermill at South Voe, Dunrossness, Shetland. The remains of a second mill are visible in the background.

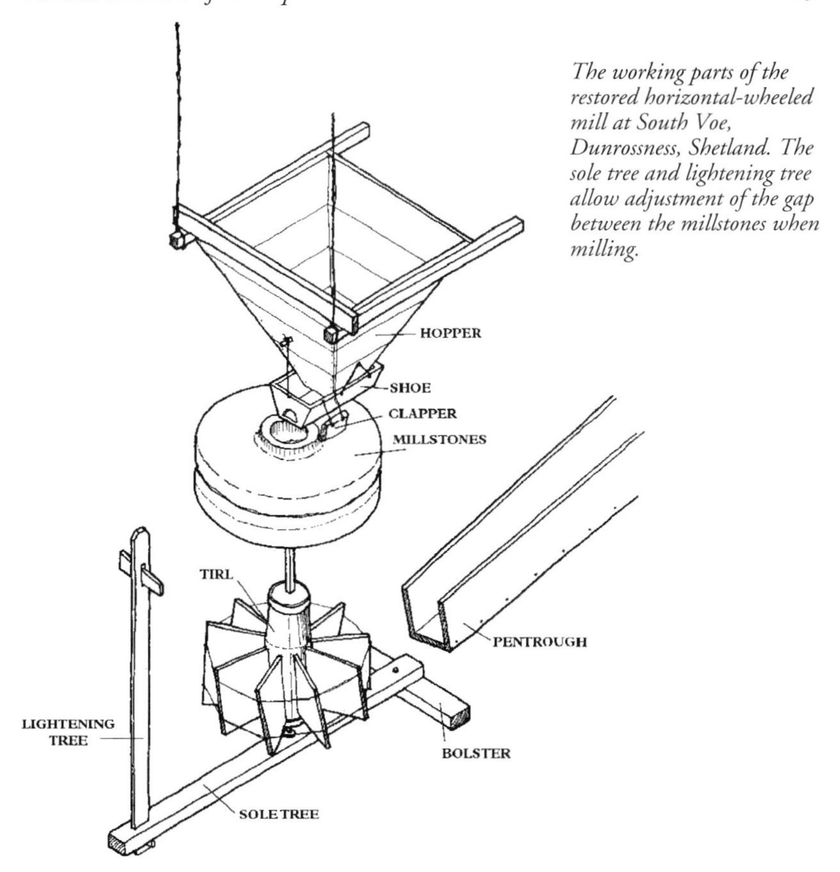

The working parts of the restored horizontal-wheeled mill at South Voe, Dunrossness, Shetland. The sole tree and lightening tree allow adjustment of the gap between the millstones when milling.

large number found in Ireland, there is both documentary and archaeological evidence from the Isle of Man, south-west Scotland, Kintyre, the Hebrides, northern Scotland, Orkney and Shetland. It has been suggested that these mills were a relatively modern introduction into the Western and Northern Isles, suited to both their geography and their economy. However, the excavation of the stone-built underhouse and watercourses of an eleventh- or twelfth-century horizontal-wheeled mill at Orphir, Orkney, in 1991, as well as some earlier finds on Shetland, calls for a re-examination of the evidence. A small millstone found at a Viking midden site on the northern island of Unst, now on display in Lerwick Museum, is probably from an early watermill. Although it is only the size of a quern stone, it has deep recesses cut for a *rynd* and was undoubtedly power-driven as it has no hole for a handle. Similarly a paddle, carved from spruce, was discovered during peat-cutting on Yell in the early 1950s. Although not typical of the flat paddles found on surviving horizontal waterwheels on Shetland, it is similar in form to the early Irish and other medieval examples that have been found elsewhere.

The restored horizontal-wheeled mill at Siabost, Lewis, Outer Hebrides. The stone-lined lade feeds water to the wheel down a timber pentrough.

While horizontal-wheeled mills certainly existed on Orkney, they survived only in isolated districts into modern times and seem to have been permitted only for the owner's own use. Mills with vertical waterwheels were built to serve the milling needs of the larger areas under the feudal mill system introduced by Lord Robert Stewart in the sixteenth century. The sole surviving example of a horizontal-wheeled mill on Orkney, Click Mill at Dounby, was not built until about 1820 and the building is significantly larger than most other remains. On Lewis in the Outer Hebrides there are considerable remains of horizontal-wheeled mills in certain areas, although it is notable that where large meal mills with vertical waterwheels and more than one pair of millstones were built in the nineteenth century the remains of the little mills appear to have been deliberately cleared away. On Shetland the situation was different, each crofter being his own master, owning his croft, boat and mill, or at least a share of the last. Dozens of small horizontal-wheeled mills, or their sites, can still be found but, because of the scarcity of timber, much of which came from driftwood and wrecks, the only remains tend to be low rubble-stone walls and traces of the watercourses. Because of their simple construction using locally available materials, it is almost impossible to date them. On Shetland and Lewis mills were usually located on burns that ran from lochs towards the sea and as many as nine could be found on one watercourse, each with its own stone-lined *lade* and sluices. It is probable that some of these sites are of considerable age and may date back to the Norse settlement of the

Horizontal-wheeled mill at Troswick, Shetland, showing the millstones and meal trough. The top of the lightening tree is visible to the left.

islands, for even the simplest waterworks took time and effort to construct and, once established, were likely to be used as long as the demand for water power for milling grain continued.

MEDIEVAL MILLS

It is through research into historical documents, particularly manorial accounts, along with archaeological excavation and a more perceptive analysis of what remains in and around surviving mills that are known to occupy ancient sites, that a fuller picture of the use of water power in the medieval period is gradually being pieced together. Domesday Book reflects the state of England at the time of the Norman Conquest and shows that by 1066 the Anglo-Saxons had a well-established grain-milling system. Several different estimates of the total number of mills recorded by the survey of 1086 have been made but in *Domesday England* Professor Darby has calculated a total of about 6082 mills. These are distributed throughout eastern and southern England, with numbers declining in the south-west peninsula west of the river Exe, towards the Welsh borders, and in north-west and north-east England, beyond the rivers Trent and Humber. It is not possible to calculate an exact number as some mills were held by two or more people and some by adjoining villages but on average there was one mill to every forty households over the whole of Domesday England.

While the survey records mills under the manors they served, it does not give exact locations, or any information other than ownership and value. These values ranged from a few pence to several pounds, although some mills returned nothing and others are noted as having gone since 1066. Some were stated to be unrented, others to grind their own grain, and a mill at Tavistock, Devon, was recorded as serving the needs of the abbey there. A few mills are referred to as winter mills as they did not have adequate water to operate during the summer. While the majority of returns are given monetary values, some were in kind, in eels, honey, salt, malt, rye or other grain, for example. Four mills on three holdings at Lexworthy, Somerset, paid a total of six blooms (bars) of iron, although this should not be taken as direct evidence that they were used for ironworking. Norfolk had the greatest number of mills (538), while Cornwall had the lowest number of six, well below the average, with only one mill to every thousand households. Cornwall was surveyed on the same circuit as Devon, Dorset and Somerset, however, each of which had a considerably higher proportion of mills to every thousand households (seven, 43 and 34 respectively). It does not seem tenable that the watermill, first recorded in Kent in the eighth century, should have taken three hundred years to reach the south-west or even the north of England, as has been suggested. It is more likely that the local economy and diet had an effect on the need for water-powered mills and perhaps the milling requirement was adequately served by the use of hand-driven querns and mortars.

The greatest density of Domesday mills was in central-southern England. Groups or clusters of mills are found in many locations but it appears that not all feudal tenants had a mill close at hand, particularly in areas where water supply was poor or intermittent because of topography. Where good natural

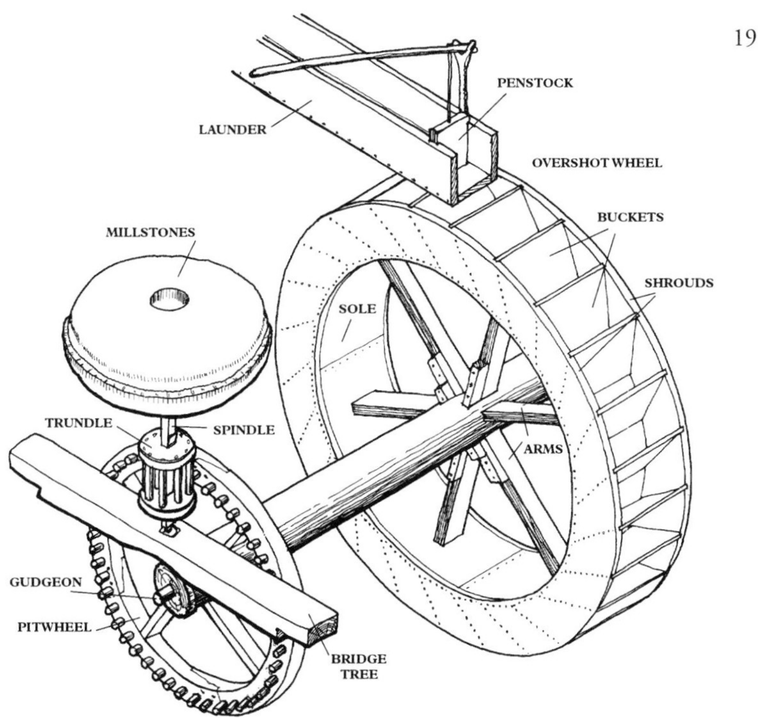

Conjectural reconstruction of the working parts of a medieval watermill.

watercourses were available, mills were often found in large numbers, such as in the Wylye valley in Wiltshire, where some thirty mills were sited along a 10 mile (16 km) stretch of the river. Where more than one mill was recorded at apparently the same location, it is probable that more than one waterwheel was used. There is no evidence that Domesday waterwheels, whether vertical or horizontal, drove more than a single set of millstones, and the term 'mill', therefore, described one milling unit, comprising a waterwheel and a single pair of millstones. It is quite likely that two or more such units were sometimes housed in a single building and were served by the same watercourse. In later medieval documents references to 'two mills under one roof' frequently occur, and there are many surviving examples of this arrangement.

In *The Mills of Medieval England* Dr Richard Holt lays emphasis on the evidence that mills were built for the benefit of the lord, not the peasant, and the system by which the unfree tenants of a manor were compelled to have their grain milled at the lord's mill and pay a toll in kind was an important and regular source of income. Although on average mills were unlikely to have contributed more than ten per cent of the whole income of an estate, it was obviously worthwhile for landlords to build them and to enforce the compulsory suit of mill, with the financial advantage that this had to the feudal economy. Free tenants did use the lord's mill, although they were at liberty to use hand mills or have their grain milled outside the manor, but the service still had to be paid for, usually in kind. For those compelled by law to use the manorial mill, fines were imposed if they did not, so whether the lord's mill received their custom or not, some form of income was guaranteed.

Although the historical record becomes increasingly significant from the end of the twelfth century, of the mills and watercourses themselves only a few sites have been found by excavation. At West Cotton, Raunds, Northamptonshire, three successive leats about 12 feet (3.6 metres) wide and 2 feet 6 inches (0.8 metre) deep have been identified. They originated in the ninth or tenth century to take water from the river Nene. The latest served a watermill of eleventh- or twelfth-century date, which probably had a vertical waterwheel. The mill structure spanned the leat and may be contemporary with the development of a range of twelfth-century stone buildings. At Hemington Fields, Castle Donington, Leicestershire, at the confluence of the Trent, Derwent and Soar, a twelfth-century mill dam and timber waterwheel race were found during gravel extraction in 1985. The dam, which probably originated as a fish weir, was constructed of two rows of oak posts with hazel wattles between them supporting a core of stone, gravel and brushwood. It was used to direct water into a mill race, and the oak structure, which had contained one wide or two narrow waterwheels side by side, has been dated to about 1140. The waterwheel was probably undershot or *breastshot* and, from the shape of the timber race, was about 7 feet 6 inches (2.3 metres) in diameter and up to 4 feet (1.2 metres) wide. A great number of millstone fragments, both used and unused, were also found.

The use of timber structures to house waterwheels appears to have been common practice where the wheel race or pit was not formed in solid rock or masonry. Wattles were also widely used, both to retain stones and other materials to form dams and to floor wheel races, to prevent the water from scouring as it left the wheel. At Batsford Mill, East Warbleton, East Sussex, the remains of an *overshot wheel* of probably fourteenth-century date were excavated in 1978. The waterwheel, of which almost half survived, was located in a wheelpit framed and boarded in oak. About twenty oak gear *cogs* were also found, but no mill building. It is thought that the wheel powered a corn or *fulling mill*. Parts of medieval timber gears have also been found by archaeologists excavating waterlogged mill sites, the most significant being the remains of a face gear with thirty-six cogs and a smaller pinion, found buried in silt at the Town Mill site in Reading, Berkshire, in 1998. The typical drive in a medieval watermill was from a *pitwheel*, a face gear mounted on the *wheelshaft*, meshing with a *trundle*, a *lantern pinion* attached to the stone spindle. The millstones would have been supported on a stout timber *hurst* and driven from below.

Weirs and mills were constructed, for preference, in the summer months, when water levels were low and work could be carried out more comfortably. In an eleventh-century document outlining the duties of the reeve, a manorial official, it was stated that 'In May and June and July, in summer, one may . . . construct a fish-weir and a mill.' Watercourses were often kept clear of

An early-sixteenth-century carving of a windmill on a church bench-end at Bishop's Lydeard, Somerset. Part of the trestle is buried in a mound; the miller, with his packhorse and a measure for taking his toll, stands by.

overgrowth and silt by manorial tenants as part of their duty to the landlord. In 1355 the manorial court at Abbots Langley, Hertfordshire, declared that all tenants, whether major or minor, should dredge and repair any damage to the mill-pond when necessary. A nineteenth-century stone- and concrete-faced weir on the river Aire at Fleet Mills, Oulton, West Yorkshire, where mills have been recorded from the thirteenth century, was found to contain a succession of timber-built structures on different alignments. Each phase was of timber box-frame construction with pegged joints, the back of the weir being made of vertical boards held inside the framework by carefully laid rubble-stone.

Water was sometimes brought a considerable distance to serve a mill, both to give a reliable supply and an adequate head or fall. At St Bees, Cumbria, the Benedictine priory was situated in the valley of the Pow Beck, a sluggish stream with a small gradient. In the mid thirteenth century the monks were given the right to take water from the Rottington Beck in the next valley, and a leat some three-quarters of a mile (1.3 km) long was dug along a contour line to supply a mill situated close to the priory. Like many such watercourses which were built with considerable effort and expense, the St Bees leat remained in use for over six hundred years.

The introduction of the windmill

Before the end of the twelfth century a new type of mill powered by wind appeared in western Europe. The origins and early history of the windmill, like those of the watermill, are surrounded by controversy, but it seems that the form most usually found, with its sails rotating in an almost vertical plane, was first built during the last quarter of the twelfth century. A tentative connection between the crusades and windmills has often been made, but the theory that windmills were first seen in the East and the idea brought back to the West by returning crusaders still remains to be proved. Between the end of the second crusade in 1149 and the beginning of the third some forty years later, however, windmills were being built on both sides of the English Channel.

The earliest authentic references in Britain occur at Amberley, West Sussex, where a windmill had been built since 1180 by Bishop Seffrid II of Winchester, and at Weedley, East Yorkshire, where a mill was standing in 1185. Claims for earlier references from the first half of the twelfth century have not been upheld by some medieval scholars because they do not specifically refer to mills powered by wind. By the close of the century, however, nearly two dozen definite references are known, all of them to windmills located east of a line between the Solent and the Tyne. Documentary evidence indicates that windmills spread throughout England during the thirteenth century, but their early adoption was most noticeable in areas where water power was not so reliable, such as East Anglia and the edges of the Fens. Dr John Langdon has estimated that there were probably over twelve thousand mills in England by the beginning of the fourteenth century, of which one-third were windmills. They were built to supplement watermills, not to supplant them, and were undoubtedly seen by landlords as a useful source of income, even though building and maintenance costs were high. The windmill at Weedley stood on land belonging to the Knights Templar and was probably connected with their need to raise money for the crusades, for their manors were chiefly of value to support their work in the East. During the reign of Henry II (1154–89) social and economic stability returned to England after civil war, and in the 1180s a period of agricultural prosperity began, accompanied by a sharp rise in prices, which surely encouraged those with milling monopolies to invest in the new invention. It is also significant that in about 1180 dramatic new ideas came into English timber building generally and both timber-framing and the saw were reintroduced, probably from France.

The earliest form of windmill found in England is the *post mill*, a triumph of medieval carpentry in which the sails, gearing and millstones are carried by a timber-framed structure that revolves about the head of a vertical post, so that the sails can be turned to face the wind from whichever direction it blows. While it may seem extraordinary to construct such a machine, the practices of building movable timber structures, raising heavy weights and using mechanical force were prevalent in the twelfth century. It was customary for leading crusaders to be accompanied by engineers, usually carpenters who specialised in siege works and engines, and it is possible that some of the leading military engineers of the time were involved with making the first post mills stand up and work. In medieval documents mill builders are usually referred to as carpenters and it is not until later in the fourteenth century that the name 'millwright' first appears.

The remains of medieval windmills that have been discovered by archaeology

Excavation of a fourteenth-century windmill mound near Bridgwater, Somerset, in 1971.

comprise parts of the substructure or *trestle*, usually buried in an earth mound. Many mill mounds have been excavated, often mistaken for prehistoric barrows, and indeed existing tumuli were occasionally used by windmill builders to give their mills extra height, as at Rodmarton, Gloucestershire, where a neolithic long barrow is still known as Windmill Tump. Some early archaeological reports fail to conceal the disappointment of the excavators in finding substantial oak timbers rather than treasure or human remains. Dating mounds and timbers has usually relied on associated evidence such as pottery, although the *crosstrees* found in 1977 at Great Linford, Buckinghamshire, have been

Part of the crosstree and one quarterbar of a post mill unearthed during the excavation of the mound near Bridgwater, Somerset, in 1971.

radiocarbon-dated to 1220 ± 80. At its earliest, this puts these timbers into the first age of windmill building. Buried trestle timbers have been found packed in clay or stone, intended to prevent or slow down decay, and the crosstrees are sometimes supported on natural stone or masonry walls. Excavated trestles are usually incomplete; *quarterbars* have occasionally been found, but rarely the post itself, which was a massive timber and would probably have been robbed out for reuse.

The reason for burying the substructure of post mills in mounds was primarily to give stability, and it appears to have been common practice in the Middle Ages. At Newborough, Anglesey, in 1303 David the Digger was paid for making the foundation of the windmill, and in 1366, when a windmill was repaired at Wymondley, Hitchin, Hertfordshire, the ground around the post was sunk as far as the foot of the post, then the soil replaced and raised to form a mound. At Patrington, East Yorkshire, in 1426 a new post was raised and set in position by ramming after the old mill had been blown down. It seems that the post was not supported by crosstrees and quarterbars in all cases, but merely embedded in a mound. While some information about the substructure of medieval post mills can be gleaned from the remains of buried trestles, knowledge of their super-structure and working parts relies on building accounts and contemporary illustrations.

A number of early building accounts survive which provide insights into the cost of building and maintaining medieval mills. At Walton, Somerset, a post mill built by Glastonbury Abbey in 1342–3 cost £11 12s 11d, of which half was spent on the wages of the carpenter and the smith

Post-mill framing, South Normanton, Derbyshire. Although from a later period, the framing shows the basic superstructure of a post mill.

Layout of a post mill driving a single pair of stones, from Abraham Rees's 'Cyclopaedia', 1819. The sails, STV, are fixed to the poll end, A, of the timber windshaft, B. The clasp arm brakewheel, C, meshes with a trundle or lantern pinion, E, which turns the upper millstone by the spindle, G. Grain is fed into the millstones, located in the circular case, K, from the hopper, P, and the ground meal emerges from the spout at O, to be collected in the ark or bin, X. The gap between the millstones, and thus the texture of the meal, is controlled by adjusting the beam, R, by a system of levers and a counterweight.

and half on materials. Of the whole cost, over eleven per cent was spent on a pair of millstones, and a pair of these was commonly the most expensive single item in a mill. Over the next few years the maintenance of the windmill at Walton accounted for about a quarter of its initial building cost each year. Some technical information can also be gleaned from such accounts. At Walton, alder stakes were gathered from the moor to make the *sail bars*, the sails themselves being made of elm and set with canvas, presumably woven in and out of the bars as is shown on some contemporary illustrations. New canvas for sails was a frequently recurring expense, as at Turweston, Buckinghamshire, in 1302–3, when 51 ells (about 58 metres) were bought, along with cords for the same, for 14s 3d. The

A medieval watermill depicted in early sixteenth-century stained glass in Thaxted church, Essex.

iron or steel spindle, by which the millstones were driven in both windmills and watermills, was subject to tremendous torque, a twisting force, and was regularly in need of repair and strengthening by a smith. Other shafting and gearing was of timber, with iron hoops and bands to strengthen it, and hundreds of nails were used in the construction of mills. Iron pins were driven into the ends of shafts to form bearing pintles, and brass for bearings is sometimes mentioned. Archaeological evidence confirms that blocks of hard stone or pebbles were also used for iron journals to run against, as they had been since Roman times. One item frequently mentioned is a lock for the mill door or for the meal ark, the chest in which the ground grain was collected as it came from the millstones, which again emphasises the value of grain and meal to feudal landlords. Frequent thefts from mills are known from manorial court records and were often met with heavy fines.

From the mid thirteenth century windmills are illustrated in a variety of contexts, particularly in manuscripts, and in paintings and carvings in churches; by comparison medieval depictions of watermills are rare. The majority of illustrations that have survived are of post mills and while they are undoubtedly stylised, they do provide some useful information. Four sails appear to be the rule, although a six-sailed mill was referred to at Framlingham, Suffolk, in 1279. The working parts of the post mill were reached by a set of steps from ground or mound level and the whole was turned to the wind by a *tailpole*, a timber beam projecting from beneath the rear of the body.

Because of their construction and often exposed locations, post mills have always been vulnerable to storm damage, and even burying the substructure and post in a mound could not prevent them from being blown over, particularly as timber decays faster at ground level. An alternative was to build a masonry tower with a timber cap, which could be turned to the wind, carrying the sails and *windshaft*. The earliest known reference to a *tower mill* in England is to that built at Dover Castle in 1294, when a new stone windmill, erected by the king's orders, replaced an earlier mill. The masonry was by Nicholas of Aynho and the carpentry by John of Harting; the mill cost £36 6s 11d, over three times more than the Walton post mill built some fifty years later. Perhaps the oldest windmill tower to survive in England is at Burton Dassett, Warwickshire, which may date to before 1367, when a ruined windmill called 'le Stonmilne' was held by Sir John de Sudeley. Certainly by the fifteenth century tower mills appear in illustrations, particularly wall paintings, as at Cottered in Hertfordshire and, formerly, Broadchalke in Wiltshire and Whimple in Devon. The form of these early towers was cylindrical, or almost so, with a conical or gable-shaped cap and four short sails, on which cloths were set from ground level. The sails were turned to face the wind by means of a tailpole that extended from the rear of the cap to the ground. Internally they contained only a single pair of millstones, as did the post mills, driven directly from a headwheel mounted on the windshaft and driving the millstones through a trundle gear from above. It is not known when the brake that acted around the circumference of the headwheel was first introduced into windmills, and it is likely that early mills were slowed and stopped by turning the mill to put the sails out of the wind.

Remains of a probable medieval stone windmill tower at Burton Dassett, Warwickshire.

Castle and abbey mills

Because of their siting, exposed to the action of running water or to the wind, and their mechanical function, mills have always required frequent repair and rebuilding. While many of those that remain undoubtedly occupy ancient sites, the standing structures were probably rebuilt comparatively recently and working machinery still to be found in mills is almost certainly less than three hundred years old. A few mill buildings do survive from the later Middle Ages, however, associated with some of the best-known monuments of that period – castles and abbeys. Although built for totally different reasons, they have a common link in that they were primarily intended for self-containment: castles to keep out the enemy, abbeys the material world. As the main function of these buildings was superseded, for social and political reasons, by the end of the medieval period, the remains that survive can sometimes represent that age quite clearly.

Mills were built as an essential part of fortifications, along with granaries and bakeries, from soon after the Norman Conquest, enabling castles to be self-sufficient during times of war and siege. Accounts from castle-building and repair work often include mills, usually watermills as the use of water as a defensive barrier encouraged its control and use for driving waterwheels. Mills built outside the bailey or castle walls, however, would have been vulnerable to attack. Carlisle Castle, Cumbria, acquired a granary and a *horse mill* in 1194, which enabled the garrison to feed itself should it be cut off from its mills on the river below. At Leeds Castle, Kent, where the mill was located in a late thirteenth-century fortified tower on the outer barbican and fed by water from the moat, there are frequent references to its repair and the replacement of timbers. The waterwheel was breastshot and about 10 feet (3 metres) in diameter, and the mill was in use up to the sixteenth century. At Warwick the mill is sited on the river Avon below the south-west wall of the castle and probably dates back to the middle of the twelfth century, since when it has operated almost continuously. It was used to grind corn for the garrison during the Civil War and from that time it also pumped water for the domestic needs of the castle. By the end of the nineteenth century water power was also used to generate electricity, giving this site a long history of self-sufficiency. At the Tower of London, Edward I had mills constructed at each mouth of the moat in about 1275, one towards the city and one or perhaps two towards St Katherine's. They were probably largely dependent on tidal power, water from the Thames being admitted by the sluices that were used to fill the moat. These mills were apparently unsuccessful because of silting problems, and those towards St Katherine's were obliterated by an extension of the fortifications in the early 1290s, although the beech piles off which they were built were found during excavations in the moat in 1997.

Perhaps the most substantial remains of a mill within the outer ward of a castle survive at Caerphilly in South Wales. Here the watermill was constructed in the thirteenth century, its wheel fed by water from the inner moat. It was probably overshot, working on a head of about 10–12 feet (3–4 metres).

The consolidated walls of the medieval mill at Caerphilly Castle, South Wales. The tailrace arch is in the centre of the picture, the spillway in shadow to the right.

The consolidated walls of the mill with its wheelpit and the paved spillway channel to the south can still be seen, although several phases of building are apparent as it was in use for four hundred years. Some castle mills survive that undoubtedly occupy ancient sites but have been substantially rebuilt in recent times, as at Newport, Pembrokeshire, where an eighteenth- or early nineteenth-century building exists close to the now domesticated ruins of the castle. Water for the wheel was originally let through a sluice from the castle moat, described in 1594 as 'a faire cleere springe of sweete runing water'. At Carew, also in Pembrokeshire, the castle mill was latterly a tide mill, worked by water being impounded in a large pond at high tide and released on to waterwheels as the tide ebbed. The preserved building, waterwheels and machinery are all relatively modern for, as is often the case, utilitarian buildings such as mills retained their value and usefulness long after the grander buildings they served had become obsolete.

The rise of monasticism in the fifth century AD was to have a dramatic effect on the use of water and wind power in Britain later in the Middle Ages, even though the original intention of monastic communities was to provide a disciplined, ascetic existence given to prayer and humility, away from the distractions of the world. The Rule of St Benedict, first laid down in the sixth century, stated that a monastery should contain every necessity of life, such as water, a mill, a bakehouse and so on, to avoid the need for the monks to go out of its bounds. Bread was both a vital part of the monk's diet and essential to the celebration of mass, and some labour-saving method of producing meal and flour was thus important. A good water supply was necessary to serve the kitchens, wash-houses and lavatories used by the community, as well as fish-ponds and mills, and many monasteries had elaborate drainage systems. Although some religious houses were long-established by the time of the Norman Conquest, many more were founded during a revival in the twelfth century, particularly by new orders such as the Cistercians, who sought to re-establish the simplicity and austerity of the early Benedictines. While they encouraged manual labour, the use of water power to perform tedious tasks gave the brethren more time for prayer and worship. The early customs and statutes of the order emphasised that ideally the monks should live on the produce of their own labour, so mills were permissible only for the community's internal use and were often sited within or close to the precinct. Many monasteries, however, were given mills or the income from them by benefactors and the gap between theory and practice became a significant one.

Part of the great wealth that monasteries accumulated during the Middle Ages, one of the reasons for their dissolution by Henry VIII in the sixteenth century, was from the profits of milling. Although some abbeys were extremely rich and held large areas of land, others were less well endowed. The Cistercian nunnery at Baysdale, North Yorkshire, for example, was worth less than £28 a year at the time of the dissolution, in contrast to Glastonbury

Fountains Abbey Mill, North Yorkshire: the downstream face of the Cistercian mill, the lower walls of which date from the twelfth century.

Fountains Abbey Mill from upstream. The millpond was backfilled in the 1920s, leaving only a narrow leat to feed a turbine. The sawmill waterwheel house is to the right.

Abbey, Somerset, which was one of the wealthiest, with an annual income in excess of £3000 in 1535. But even the poor nunnery had its own mill, described at the time of the suppression as an overshot mill hard by the gate, 20 feet (6 metres) long and 14 feet (4.3 metres) broad, with stone walls, part boarded and part covered with thatch, although it was then in decay 'so that the seid mylne goith not'.

While evidence of watermills is to be found at several monastic sites, the finest survivor is that in the outer court at Fountains Abbey, North Yorkshire. The earliest masonry probably dates from 1135–46, the first stone building phase of the abbey, although the mill was substantially reconstructed shortly after and further altered early in the thirteenth century. The main building is about 72 feet (22 metres) long by 33 feet (10 metres) wide and was probably originally built across the course of the river Skell, with a central wheelpit that housed two undershot wheels. There is archaeological evidence of a stone dam that was built to impound a working head of water immediately on the upstream side of the mill, and there have been a number of subsequent alterations to the water supply, all of which play an important part in understanding the development of this site. The size of the building is considerably larger than that of other known medieval mills and it is likely that it may also have served as a granary and perhaps for other industrial purposes. In 1540 it was referred to as two water corn mills under one roof, worth 40s (£2) a year. Like some castle mills, the building has survived because the mill remained in

The Old Malthouse, Abbotsbury, Dorset: the downstream face of the fourteenth-century monastic watermill.

use into the twentieth century, latterly using a waterwheel to power the estate sawmill and a *water turbine* to generate electricity. At Abbotsbury, Dorset, a building known as the Old Malthouse has been identified as the fourteenth-century watermill of the Benedictine abbey. When the wheel chamber was excavated in 1984 it provided evidence of two waterwheels located side by side in the middle of the building. They were probably overshot and up to about 13 feet (4 metres) in diameter. Circular scratch marks on the wheelpit walls, stone bearings and fragments of millstones have all given information about this mill, which survived the dissolution of the abbey in 1539 and finally went out of use in the eighteenth century. While it was undoubtedly the corn mill of the abbey, its more recent name implies that it may have once contained a malt kiln or was used also to grind malt for brewing.

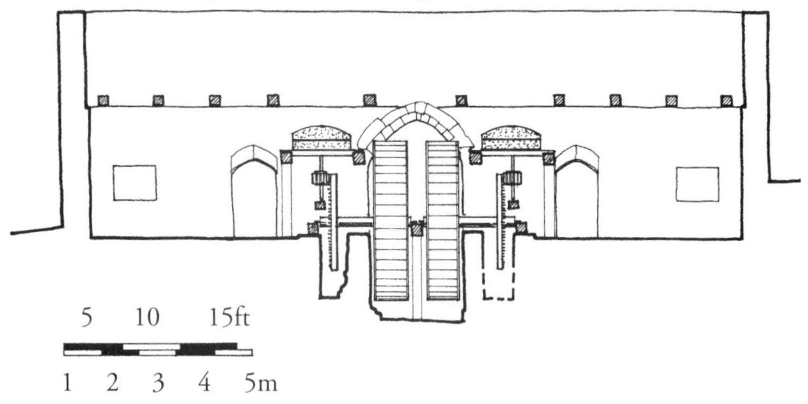

Section through the Old Malthouse, Abbotsbury, showing the positions of the waterwheels and millstones, 'two mills under one roof' (after Alan Graham).

An early-seventeenth-century carving of an undershot waterwheel driving a pair of fulling stocks, from the Tuckers' Hall, Exeter, Devon.

Fulling mills

As well as corn and malt mills, there is evidence that many religious houses used water power for other purposes, most notably for fulling cloth, driving mills for crushing oak bark to make tannin for leather working and for forging iron. Mechanical fulling is probably of European origin, its purpose being to cleanse woven woollen cloth of superfluous grease and oil and to pound it in order to mat together loose fibres so that it was both felted and shrunk, by up to twenty per cent of its original length, making it fit for sale and use. Fulling was carried out by placing bolts of woven cloth in a container with water and some form of scouring agent, such as fuller's earth, and pounding them with stocks or stampers. Fulling *stocks* are like large wooden mallets, their handles tripped by *cams* driven by a waterwheel to give a beating action. *Stampers* are vertical timbers, lifted by cams and allowed to fall by their own weight. Stocks were used for scouring and stampers for felting or milling the cloth.

Fulling mills are first definitely recorded in England in 1185 at Temple Newsham, West Yorkshire, and Barton on Windrush, near Temple Guiting in Gloucestershire, both on lands belonging to the Knights Templar. At Kirkby on Bain, Lincolnshire, however, a fulling mill belonging to the Cistercian abbey at Kirkstead is mentioned in a charter that may date back to 1154, and at Heycroft, near Malmesbury, Wiltshire, a fulling mill may have existed as early as 1174. Fulling mills seem to have been adopted with even greater enthusiasm than windmills, which appeared at about the same time, and large numbers were built by the end of the twelfth century, particularly in areas where there was an already established cloth industry and plenty of water for power. Although the Cistercians drew much of their wealth from wool production rather than cloth manufacture, probable fulling mills have been excavated at Fountains Abbey and at Beaulieu Abbey, Hampshire. At Fountains the mill is thought to have been powered by a small undershot wheel and to have been rebuilt towards the end of the thirteenth century. During the fourteenth century the building was reused as a dyehouse, so that fulling must have taken place elsewhere, perhaps in part of the large corn-mill building, the cloth being brought back for dyeing and finishing.

A new fulling mill was built below the corn mill on the royal estates at Elcot, Marlborough, Wiltshire, between July 1237 and February 1238, the accounts referring to 'flagella et baterell', presumably stocks and stampers. In 1427 a carpenter was paid 21d for making 'le stooke' of the fulling mill of Hoggefordmull, near Wimborne, Dorset, a task which took him three and a half days. Ten years later, two millwrights agreed to build a fulling mill at Chartham, Kent, for £14 13s 4d. Although fulling mills appear to have been built close to or even adjoining corn mills, they were usually powered by a separate waterwheel. An exception apparently existed at Layham, Suffolk, where in the latter part of the fourteenth century the same wheel drove a pair of millstones and a set of stocks. As stocks and stampers could be operated by cams driven into an extended waterwheel shaft they did not need intermediate gearing and generally required less power than a pair of millstones.

Ironworking

The majority of medieval mills used for purposes other than grinding corn were for fulling but, although they were subject to compulsory suit under manorial custom, they were not as profitable as corn mills. The action of fulling stocks, which are simply power-driven hammers, had another obvious application, however, for metalworking. Waterwheels were used to drive hammers and bellows on the European mainland at least as early as the thirteenth century, while in England an early documentary reference to a hammer mill for working iron comes from Warley, West Yorkshire, in 1349. The substantial remains of what has been identified as a late-twelfth-century water-driven iron forge, however, have been excavated near the site of the Cistercian abbey at Bordesley, in the valley of the river Arrow near Redditch, Worcestershire. The first mill was destroyed by fire shortly after being built but was replaced by a similar structure that survived for about a century. Early in the fourteenth century the wheel trough and tail race were reconstructed and the mill again rebuilt, although the earlier hearths were retained. In the mid fourteenth century the building ceased to be used for metalworking and the site subsequently went out of use, probably because of silting downstream. Parts of a waterwheel, stone bearings, a cam wheel and timber pegs (probably cams for tripping a hammer) were found in addition to the timber-lined wheel race. The waterwheel was undershot, about 10 feet (3 metres) in diameter by 18 inches (0.45 metre) wide, and drove bellows for maintaining the temperatures of the hearths and a trip hammer for working the heated metal. The forge was used for the manufacture of small iron items such as nails and tenterhooks, the latter being used for holding newly fulled or dyed cloth out to dry on wooden frames. Some working of copper alloy and lead also appears to have taken

A sixteenth-century timber-lined wheelpit and the remains of an overshot waterwheel excavated at Chingley Furnace, Kent.

place. It is possible that larger items such as knives and tools were also made, and the forge probably served more than just the monastery. The parts of the waterwheel and machinery were made of oak, but the pegs were cut from apple wood, an early example of the use of such hard, close-grained timber, always much favoured by millwrights for gear cogs.

A timber-framed structure that contained the remains of a waterwheel and that was probably part of an early-fourteenth-century water-powered forge was excavated in the early 1970s at Chingley, in the Weald of Kent, an area well known for its ironworking in the post-medieval period. The waterwheel was of oak and, although the water-supply arrangements were unclear, was probably overshot. It is estimated to have been about 8 feet (2.5 metres) in diameter by 1 foot (0.3 metre) wide and capable of producing about 1$^1/_2$ horsepower (1.1 kW). This early ironworking installation appears to have been abandoned in the mid fourteenth century, although the site was subsequently reused towards the end of the sixteenth century and parts of two later waterwheels were also found.

While the use of waterwheels enabled smiths to increase their output, driving bellows that could supply a continuous draught and hammers that enabled larger pieces of ore or metal to be worked, the actual power output of medieval waterwheels was generally low and the water supply must often have been intermittent or unreliable. At Byrkeknott, an iron-smelting site in Weardale, County Durham, casual payments were still being made to a woman who worked the bellows even after a new conduit was dug for a waterwheel in 1408, and many other medieval metalworking sites have been found where water power was not used. Undoubtedly the cost of digging leats and ponds, and of building and maintaining waterwheels, was an important consideration. Where financial resources were available, as well as a good site, water power was adopted more enthusiastically. At Rievaulx Abbey in North Yorkshire, the river Rye drove a fulling mill and probably a *bark mill* for the tannery within the main precinct, while an iron forge was sited a little further downstream. A corn mill situated above and to the north of the abbey and fed by three spring-fed ponds may or may not be on the site of the original monastic mill, but there is little doubt that the monks of Rievaulx had the resources and the manpower to construct artificial watercourses and to use waterwheels to perform several different functions.

Dual-purpose mills, where corn milling and fulling, for example, were carried out under one roof, were more common after about 1350, the point when the high profits of the previous century came to an abrupt end because of the Black Death. The number of mills fell as the population decreased, for the plague which ravaged England in 1348–9 is estimated to have killed about a third of the people. There was a gradual falling off in the number of mills until about 1400, then a significant decline until the end of the fifteenth century. In Winchester, Hampshire, there were at least nine watermills by about 1200, plus four serving the religious houses of the town. By 1400 only six mills remained in use, and this number was halved over the next 150 years. After the Reformation, however, the number of mills generally began to rise and the uses to which water and wind power were put also increased significantly.

THE POST-MEDIEVAL PERIOD

The period of economic growth during Elizabethan and Stuart times, due to changes of land ownership after the dissolution of the monasteries, improvements in standards of living and a rise in the population towards the end of the sixteenth century, put increasing pressure upon natural resources, particularly on the use of watercourses and land. Agriculture flourished, and more trades adopted power-driven machinery in order to increase both the range and quantity of production and to meet a growing demand. Mill owners and millers, the latter no longer merely manorial tenants or servants, were now in a position to improve both the performance of their mills and the quality and variety of their products. Although the enforcement of the landlords' customary rights connected with the use of mills, which had become well established during the Middle Ages, caused many legal disputes during the sixteenth and seventeenth centuries, millers gradually became more competitive and were able to operate in a freer climate. Between 1640 and 1750 mill monopoly declined significantly in English market towns.

One problem linked with the increase in the size of towns was the lack of suitable sites for mills to provide for growing urban populations. By the mid sixteenth century *floating* or *boat mills* had been established on many of the great rivers of Europe, such as the Seine and the Danube. They comprised a corn mill set on a barge with an undershot waterwheel supported between the hull and an outrigger, turned by the flow of the river. The mills were moored close to the shore, with access usually across narrow plank bridges. While the history of floating mills dates back to the siege of Rome by the Ostrogoths in AD 536 and they survived in Europe into the twentieth century, in England they seem to have been a curiosity. They would have been especially suitable to supplement the milling needs of sixteenth-century London and two early attempts were made to establish them on the Thames. The first was by John Cooke of Gloucester, an enterprise that started in 1516, sponsored by the Mayor and City of London. Two mills were working by 1519 and a further two were built by the end of 1523, but they appear to have been short-lived. Towards the end of the sixteenth century another floating mill existed at Queenshithe but, according to John Stow in his *Survey of London*, 'this lasted not long without decay, such as caused the same Barges and Mill to be remooved, taken asunder, and soone forgotten'.

It was during the expansive years of the late sixteenth and early seventeenth centuries that watermills and windmills began to be developed from their medieval forms. New mills were built and many of those existing were enlarged or rebuilt using more durable materials. This makes this period a significant one to study, although little research has been published, the achievements of the seventeenth century often being overshadowed by the surge of technology of the eighteenth. The construction of mill buildings generally tended to follow local vernacular traditions, with some notable exceptions, at least until the second half of the eighteenth century. There was also a general increase in written sources that give valuable information about technical and agricultural development. One of the earliest descriptions of the different types of waterwheel, their construction and efficiency appears in John Fitzherbert's *Boke of Surveyinge*, first published in 1523. During the second half of the sixteenth century a number of significant books were printed in Europe, such as Agricola's *De Re Metallica* (1556), which vividly depicts contemporary mining and metallurgical practice, and the *Theatra Machinarum* of Besson (1579), Ramelli (1588) and Veranzio (1595), for example, which illustrate a variety of water-, wind- and animal-powered devices, as well as some complex gearing forms. While it is doubtful that these books would have had any immediate effect on established millwrighting practice in Britain, there is clear evidence that technology was advancing, partly because of the continental specialists who came to England to work in a number of crafts and trades, in particular in mining, metalworking and land drainage.

Although there was a growing interest in science and technology during the seventeenth century, there was still a fundamental difference between the reason of science and the empirical solutions of contemporary millwrighting practice. New ideas were often developed slowly and many would perhaps have been considered impractical by those responsible for building and maintaining water- and wind-powered machinery in an age when strong prejudices and interests sometimes stood in the way of innovation. An interesting diversion was the *horizontal windmill*, in which the sails rotated in a horizontal plane, directly driving a vertical shaft down through a tower or mill building. Several forms are illustrated in the continental machine books, including five in Veranzio's (1595), and a number of ingenious examples were proposed,

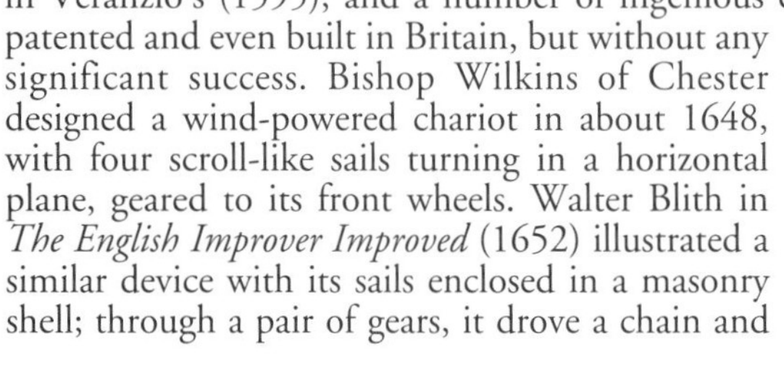

patented and even built in Britain, but without any significant success. Bishop Wilkins of Chester designed a wind-powered chariot in about 1648, with four scroll-like sails turning in a horizontal plane, geared to its front wheels. Walter Blith in *The English Improver Improved* (1652) illustrated a similar device with its sails enclosed in a masonry shell; through a pair of gears, it drove a chain and

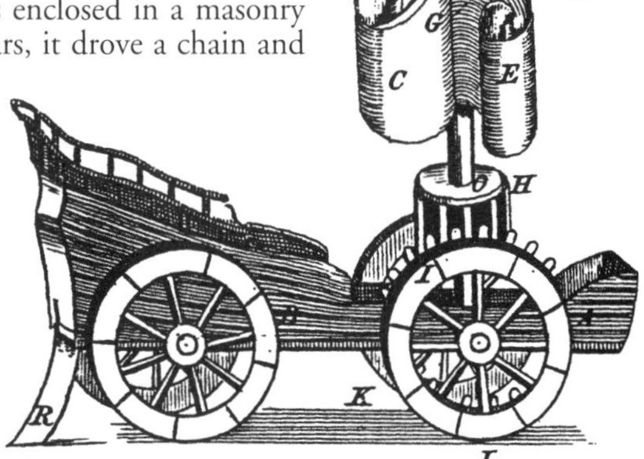

Bishop Wilkins's wind-powered chariot, from W. Emerson's 'Principles of Mechanics', 1758. The sails, CDEF, turn the trundle, H, which is geared to cogs on the front wheel, IL. R is a rudder by which the vehicle is steered.

Remains of a watermill with a single pair of millstones at Castell Howell, Carmarthenshire. Although this has early-nineteenth-century iron gearing, the form and layout are considerably older.

bucket pump to raise water from a well. In 1683 the prominent natural philosopher Dr Robert Hooke communicated a design for a horizontal windmill to the Royal Society (founded in 1660) as an 'easy way of producing a circular Motion below, without the Help of Trundles or Cog-wheels, which are both a great Impediment to its Motion, and do wear, and often need Repair.' His windmill, of which little more is known, was considered by his contemporary John Aubrey to be 'of great use for draining of grounds'.

Probably the most significant historical event of the seventeenth century was the Civil War and amongst its landmarks and victims were mills. The first battle was fought at Edgehill, Warwickshire, in 1642 and tradition relates that Charles I watched his defeat from the post mill that stood there. The town of Lyme Regis, Dorset, was deliberately fired by the Royalists in 1644 and the town mill was listed among the buildings destroyed. When archaeological work was being undertaken in 1995–6, before restoration of the mill, a layer of burnt material dating from the mid seventeenth century was found. In June 1645 Charles I witnessed his defeat from a windmill overlooking the battlefield at Naseby, Northamptonshire, and a small open-trestle post mill raised on a circular masonry base features in contemporary illustrations of the battle. The mill was finally destroyed by fire in about 1732.

Corn milling

Before the development of gearing, the only way to increase the capacity and output of a corn mill was to add a second waterwheel, along with a second pair of gears and millstones. There are many examples of mills with two waterwheels and, to make good use of both site and water supply, the wheels could be located in a number of different positions – side by side, at opposite ends of a building or in line – and they could be enclosed or outside the mill building. The watermill at Nether Alderley, Cheshire, was referred to as newly built in 1591 and its sandstone walls and fine timbered roof date from that time. The stream running through the site was small so the Elizabethan millwrights built a dam, of which the rear wall of the mill forms part, across the valley, to form a substantial pond. The position of the overshot waterwheels is unusual in that they are in line within the same building but on different levels, the lower wheel being fed by the tail water of the upper. There is evidence for a third waterwheel at the lower level and it is likely that each wheel originally drove only a single pair of millstones. The surviving *spurwheel drives*, formerly to two pairs of stones from each waterwheel, are of more recent date.

When waterwheels were placed side by side in the middle of a building, some arrangement was usually made for a spillway channel or channels, to allow water to escape when the mill was not at work or water levels were high. The wheel-house of one such *double mill* was excavated at Caldecotte, Bow Brickhill, Buckinghamshire, on a leat beside the river Ouzel, in 1980. The first construction phase, dated to 1670–80, included a timber-framed 'water house' containing four channels. The two outermost appear to have contained undershot waterwheels, each of which would have served a separate mill, while the central channels acted as spillways. A later construction phase, dated between 1720 and 1750, was similarly interpreted, suggesting the destruction and rebuilding of the mills, which finally went out of use about

Clasp-arm timber overshot waterwheel, Arden Mill, North Yorkshire. Note the simple construction and the use of nails.

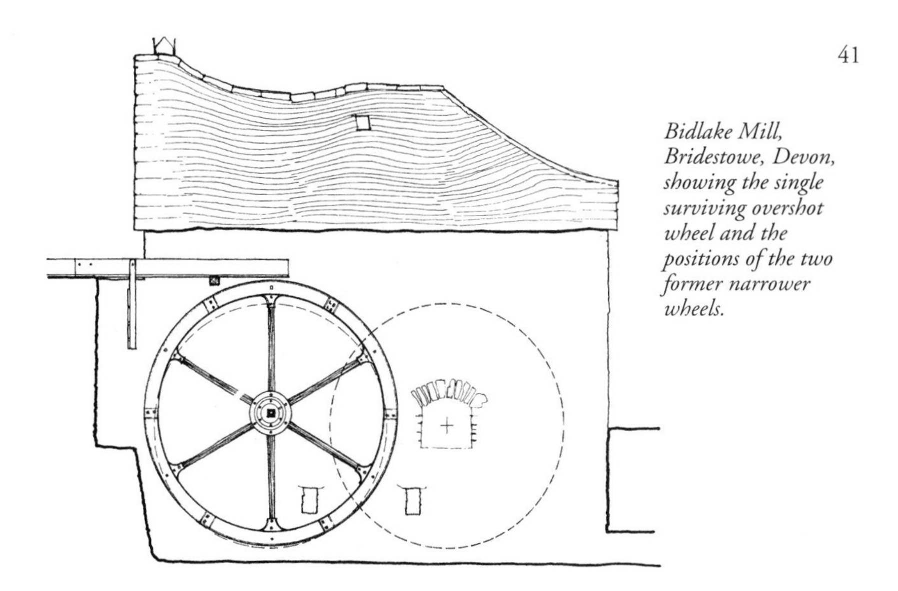

Bidlake Mill, Bridestowe, Devon, showing the single surviving overshot wheel and the positions of the two former narrower wheels.

1770. Stephen Primatt, writing in about 1667, observed that mills, although made very strong, were 'very subject to decay'. It is difficult to be precise about the working life of timber waterwheels and machinery as much depended on topography, as well as on the construction of the mill and regular maintenance. From documentary evidence it appears that an oak wheelshaft would last about eighty to a hundred years, but smaller components, particularly waterwheel floats and buckets, would require frequent replacement. In the case of Caldecotte, the low-lying nature of the site perhaps accounted for the need to rebuild the mills after only forty to fifty years.

In contrast, the positioning of two overshot waterwheels in tandem on the side of a stone building, a layout which was once relatively common in the upland areas of north-western and south-western England, has left a number of significant remains. A good example is Eskdale Mill, Boot, Cumbria, where each waterwheel drives a single pair of stones and the basic arrangement survives intact. Bidlake Mill, Bridestowe, close to the north side of Dartmoor in Devon, now has a single waterwheel geared to drive two pairs of millstones and a circular saw bench, but there is clear evidence within the building fabric that it originally had two smaller-diameter overshot waterwheels. In 1565 Henry Bidlake purchased the watercourse of the Lew river where it ran through his land for £8 in order to secure water for his mill, which was first established in the thirteenth century. The same year the 'leate was finished and the myll reedified'. Two mills were contained within the same building, each driven by a separate waterwheel. The mills were used for corn milling and *tucking* (fulling) in the early seventeenth century and were leased by a tanner in 1622, who probably made use of the existing fulling stocks for working leather. They were later included in a list of property sequestered by Parliament in about 1651 until Henry Bidlake, a Royalist, paid a fine of £300 for defending Pendennis Castle, Cornwall, 'in opposition to ye forces of Parliament'.

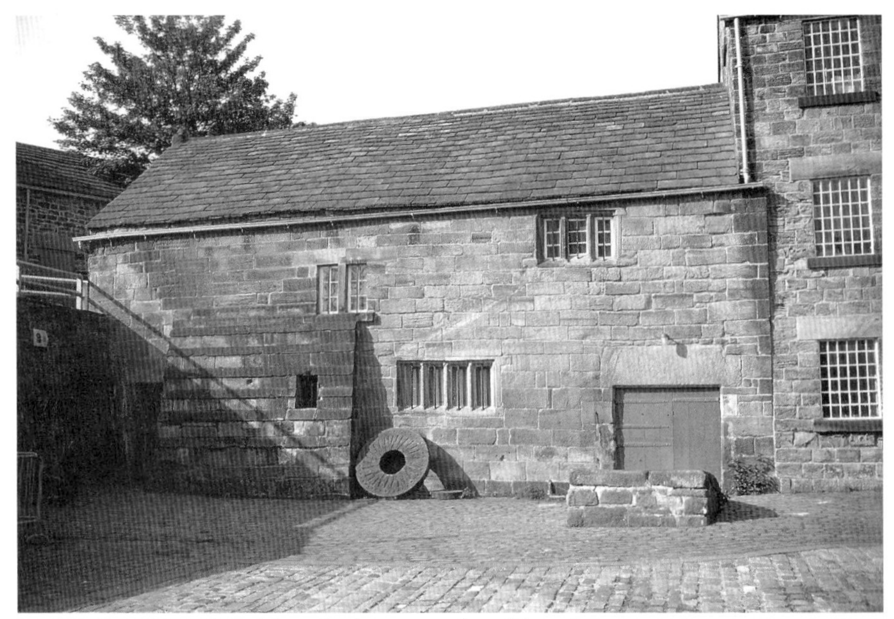

Worsbrough, South Yorkshire: the downstream face of the early-seventeenth-century watermill. The unusual millstone is probably from a bark-grinding mill and not original to the site.

Standing remains of early-seventeenth-century mills are sometimes found where stone-building traditions are strong, such as Worsbrough Mill, on the river Dove in South Yorkshire. Although documentary evidence is scant, the building has been dated to about 1625 from the evidence of the tooling on the masonry, massive lintels and sturdy king-post roof trusses. The building provides an interesting parallel with the nearby Monk Bretton Priory Mill, also of ashlar sandstone, which was rebuilt above ground by Sir William Armyne in about 1635. A hearth was recorded at Worsbrough Mill in the Hearth Tax return of 1672 and it seems likely that the miller occupied the north end of the building as his dwelling, where fireplaces still exist. The present mill house was built on an artificial terrace above the mill, probably in the mid eighteenth century, an indication of the increasing prosperity of millers. The mill has a single internal overshot waterwheel at its south end and, while the surviving gearing is of nineteenth-century date, some evidence of an earlier mechanical layout can be found within the structure.

In watermills, short horizontal shafts could be geared off one or both sides of a pitwheel, allowing drives to be taken to additional pairs of millstones through a second pair of right-angled gears. Termed *layshaft drive*, this arrangement fits well into the rectangular floor plan of a watermill and allows the millstones to be positioned in line on the hurst frame, which in some surviving mills can be seen to have been extended to carry a second pair of stones. The use of two gears on the same shaft each driving a pair of millstones is probably first represented in the *head and tail* arrangement in post mills, where the *brakewheel* drove a pair of stones in the breast of the mill and a *tail wheel* a pair located behind them in the rear. This layout appears to have been

Clasp-arm brakewheel and drive to a single pair of millstones, Stevington Post Mill, Bedfordshire. The chain drive on the right-hand side is for the sack hoist.

adopted during the seventeenth century and can be seen in many surviving post mills, such as Great Gransden, Cambridgeshire, which may date from 1612 and was certainly standing in 1674. A post mill at Great Dunmow, Essex, had two pairs of millstones and a roundhouse enclosing the trestle in 1734, the first reference to either of these features in the county. In survivals of the head and tail layout, it is not uncommon for the head or brakewheel to be of *clasp arm* construction and the tail wheel to have *compass arms.* The former provided a stronger structure, necessary for taking the band brake that acts around the circumference of the wheel to slow and stop the mill.

One of the earliest illustrations showing a development in gearing based on a working example is of a watermill at Barr Pool, Nuneaton, Warwickshire, and dates from 1723. This shows a layshaft geared off the downstream side of the pitwheel to drive a second pair of millstones. As two steps of gearing were involved, the downstream stones rotated faster and could therefore be smaller in diameter than those driven directly from the pitwheel. The use of spur gearing, with a proportionally large diameter spurwheel driving small stone nuts, also

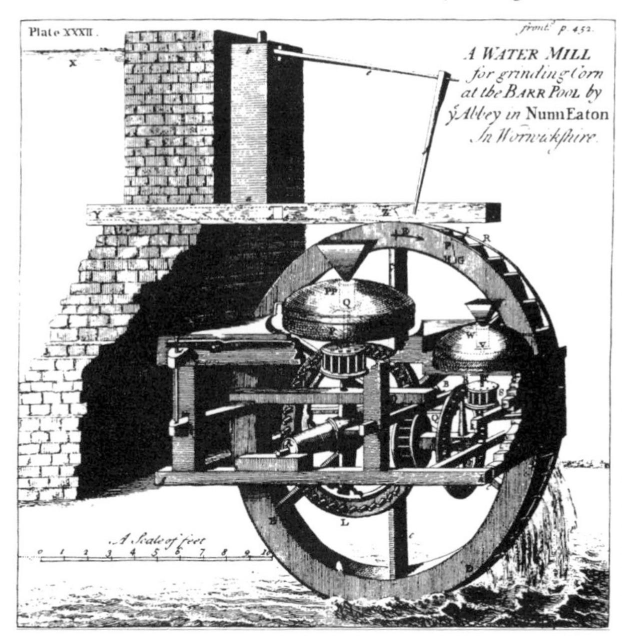

Henry Beighton's engraving showing the development of gearing to drive two pairs of millstones from one waterwheel, at Nuneaton, Warwickshire, 1723.

Layshaft drive to two pairs of millstones, through timber and iron gearing, at Cleobury North Mill, Shropshire. The waterwheel shaft is in the foreground, and part of the hurst frame and tentering control can be seen on the left.

allowed a reduction in the diameters of millstones, although this is not a hard and fast rule. The size of millstones varies with both region and period and as yet an in-depth comparative study has not been made. By the end of the seventeenth century there is an increasing number of references to double and treble mills, where more than one set of millstones driven from a single prime mover is implied, and some mill buildings from that period are of a size which supports this evidence, although most surviving machinery is much younger.

Spurwheel drive, in which a number of drives are taken by *pinions* of smaller diameter from the periphery of a horizontal gearwheel set on a vertical shaft, was perhaps developed from the horse mill. It may also first be represented in windmills as it is well suited to the circular floor plan of tower mills, where the main drive has to follow the centre of the tower in order for the cap to rotate. An early example is the unusual Chesterton Mill, Warwickshire, which has two pairs of millstones and dates to 1632. A more typical stone tower mill with two pairs of stones and spurwheel drive can be seen at Bembridge, Isle of Wight, dating from about 1701. Spurwheel drive, which was to become the dominant gearing form in British corn mills, was also adopted in watermills at about the same time; in 1737 a

All-timber spurwheel drive, Cawsey Meethe Mill, North Devon, probably late eighteenth century. The final drive to the millstones is now displaced.

Chesterton Windmill, Warwickshire, built by Sir Edward Peyto in 1632.

watermill at Westacre, Norfolk, was rebuilt to operate a spurwheel in place of two old cog wheels, with the result that it both 'saved water and performed more work'.

Two notable exceptions to the modest vernacular appearance of early-seventeenth-century corn mills are the watermill and windmill at Chesterton. Documentary evidence indicates that the watermill was built in the 1620s, and the windmill, which stands on a prominent site close by, overlooking the Fosse Way, was undoubtedly designed and built as a windmill, albeit of an unusual and unique form, in 1632. Both mills are probably the work of the landowner, Sir Edward Peyto, 'a man very skilled in the worthy art of mathematics', and the windmill may represent the earliest survival in England of the use of spurwheel drive to two pairs of stones. Although the machinery has been altered and was substantially repaired in the late 1960s, it is still of an early form, with a compass arm brakewheel and a lantern pinion *wallower*, and there is strong evidence that the windmill was built to work two pairs of millstones. Both Chesterton mills are exceptional architecturally, but they are not alone in this. At Rowlands, near Ilminster, Somerset, stands an early-seventeenth-century Ham stone and brick building of considerable architectural merit, with some fine internal oak panelling and chamfered beams. The surviving machinery represents two main phases of construction – eighteenth-century timberwork and mid-nineteenth-century ironwork. The building was almost certainly constructed as a watermill, its splendid south front forming a focal point on the original approach to the late medieval house.

Rowlands Mill, Ilminster, Somerset: the downstream façade of the early-seventeenth-century mill and house (above), and a section to show the later waterwheel and machinery that drives two pairs of millstones (below).

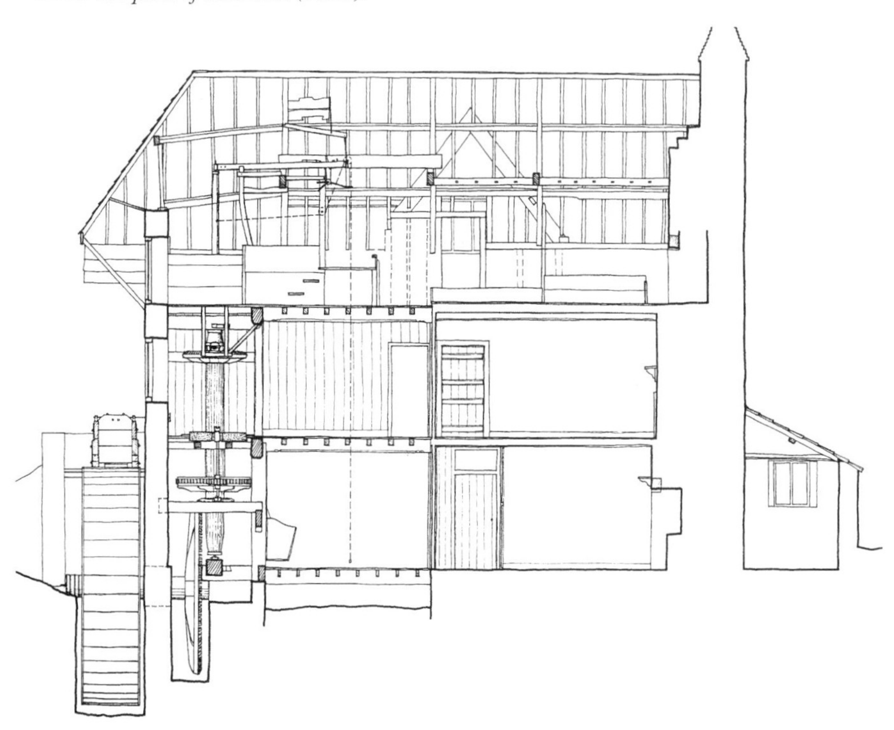

Seventeenth-century timber framing, Rossett, Wrexham. The mill roof has been raised; note how high the wheelshaft enters the mill.

Although an increasing number of watermills were built of stone, particularly where a pond or leat was embanked at a higher level to give a better head of water and allow the placement of a more powerful waterwheel with a larger diameter, timber remained the predominant and most economic material for mill building. At Stretton, Cheshire, the rear wall of the mill, which forms part of the dam of the millpond, is of sandstone blocks, while the gable ends and front of the mill are timber-framed; part of the frame bears the date 1647. At nearby Rossett Mill, situated on the river Alun, just over the border in Wales, the wheelpit and race are of ashlar stonework and, although the mill has been substantially rebuilt in brick and the roof raised, much timber framing survives within the building, and the adjoining mill house is dated 1661. The present layout of machinery further indicates that the diameter of the waterwheel has been considerably increased since the mill was built.

In East Anglia and south-eastern England the timber-building tradition remained strong. Post mills were the dominant windmill type until the early years of the nineteenth century, but a number of those surviving that are thought to date from the seventeenth century appear to have been subsequently rebuilt. Bourn Mill, Cambridgeshire, standing by 1636 and from its size and appearance considered to be of an early date, was probably rebuilt in the 1740s after being blown down. Similarly Brill, Buckinghamshire, built in about 1680, was rebuilt in 1757. Post mills were always vulnerable to storm damage but it is likely that the heavier timbers, particularly the post, would have survived to be reused in subsequent rebuildings. Daniel Defoe, in his *Tour through the Whole Island of Great Britain*, published from 1724, records the violent gales that swept across England in November and December 1703, causing much damage to buildings, including the destruction of Winstanley's Eddystone Lighthouse. It has been subsequently stated that some four hundred windmills were damaged or destroyed, but documentary evidence is inconclusive.

Although stone tower mills are known from documentary sources during the seventeenth century, they were still relatively uncommon and perhaps, like the tower mill at Dover Castle in the Middle Ages, they were considered expensive and their building relied on state patronage. The King's Windmill at Pontefract, West Yorkshire, which was described as 'the ancient stone wind mill' in the time of James I (1603–25), does not appear to have survived long after the Civil War, perhaps because of its royal connections. During the seventeenth century, tower mills became more widespread, significantly in areas where stone-building traditions were strong, such as the West Country. A stone tower dated 1571 survives on the island of Sark and two small-diameter towers standing at Easton, on the Isle of Portland, Dorset, probably date from before 1626, when they appear on a map. A tower at Landewednack, on the Lizard peninsula in Cornwall, was described as an 'old wind mill' in 1695. By the early eighteenth century windmill towers were also being built in brick. Two brick tower mills were built on the former castle mound at Devizes, Wiltshire, for crushing rape seed for oil before 1716, and at Wrottesley, Staffordshire, the tower of a former windmill has a brick inscribed 'THIS MILL WAS BUILT BY JOHN CHAMBERLAIN 1720'.

Small early-seventeenth-century stone tower mills at Easton, Portland, Dorset.

Water supply

The growth of towns during the post-medieval period led to problems of disease caused by overcrowding and inadequate water supplies. In 1582 a water-powered 'engine' was erected at the north end of London Bridge to supply Thames water to the city. The engine, one of the first in England, was built by Peter Morrice, a Dutchman or German, and comprised a waterwheel turned by the ebbing or flowing of the tide, which drove a plunger force pump. The pump lifted river water into a tower reservoir, from where it was distributed, by gravity, to conduits within a 2 mile (3 km) radius. The engine required frequent repair and maintenance and, although destroyed in the Great Fire of 1666, it was rebuilt and continued to serve the city. By the end of the seventeenth century it again required replacement, the work being undertaken by George Sorocold (fl.1660–1721), 'a man expert in making mill-work, especially for raising water to supply towns for family use', according to Daniel Defoe. Sorocold's London Bridge works of 1701 comprised four undershot waterwheels, each about 20 feet (6 metres) in diameter by 14 feet (4.3 metres) wide, driving a total of fifty-two pumps. In order to allow for the rise and fall of the tides, the waterwheels were raised and lowered by windlasses connected by gears and chains to horizontal timbers that carried the wheelshaft bearings. Sorocold was probably assisted in this design by John Hadley of West Bromwich, a 'great master of hydraulics', who had taken out a patent for a rising and falling waterwheel in 1693. Hadley had previously

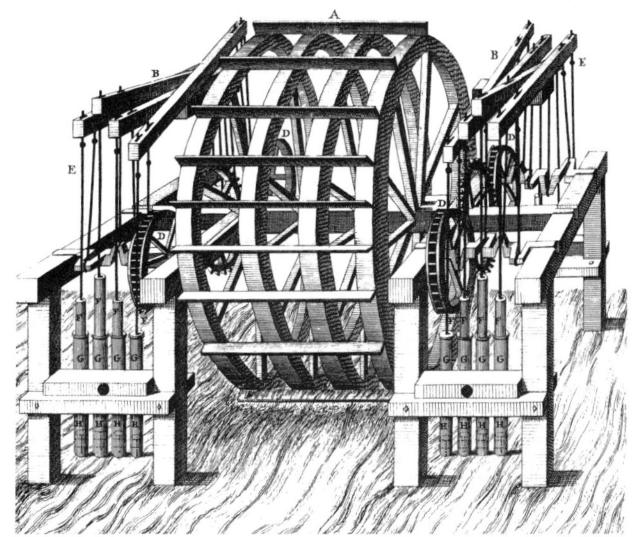

George Sorocold's water-pumping wheel driven by the Thames at London Bridge, from Stephen Switzer's 'Hydrostaticks and Hydraulics', 1729. The waterwheel, A, drives cranks through gearing on both sides. The pump rods, E, and pistons, F, are raised and lowered by the levers, B.

Sorocold's waterworks on the river Severn at Bridgnorth, Shropshire, 1776.

worked on water-supply engines for Chester and Worcester, where in 1681 his work was described as 'a wheel which gives motion to suckers and forcers, it pumps the water so high into a leaden cistern that it serves any part of the city'. He collaborated with Sorocold on Marchant's London waterworks in 1696, and he and Sorocold were probably the two millwrights of their time who, according to a contemporary, never failed what they undertook.

Much of Sorocold's work was devoted to public water supply and pumping. In 1692, on the site of a gunpowder mill by the river Derwent in Derby, he built a water engine that had a rising and falling waterwheel. As well as pumping water, it drove machinery for boring elm trees to make pipes and ground malt 'all at the same time'. He was responsible for a number of engines for urban water supplies as far afield as Exeter, King's Lynn and Newcastle upon Tyne, and his last waterworks seem to have been at Bridgnorth, Shropshire, in 1706. As early as 1694 he had received enquiries

A small overshot wheel driving pumps for domestic water supply at Morwellham Quay, Devon.

into the suitability of his water engines for pumping water from coal mines, and in 1709 he advised the use of a crank and beam pump over buckets and chains at a coal works at Alloa, Scotland, but it appears that the local millwrights were unable to execute his scheme. He was also responsible for smaller-scale pumping engines, such as that built for Sir Godfrey Copley to supply his house at Sprotborough, South Yorkshire, in 1703, the descendant of which still partly survives. Other large houses and estates also introduced water-raising machinery during this period, worked by animal or water power; Gorhambury in St Albans, Hertfordshire, had a piped water supply by 1637 and, amongst a number of references to frequent repairs to the system in the household accounts, a carpenter was paid 1s 6d in 1638 for mending the waterwheel.

The textile industry

George Sorocold was also an important figure in the early development of textile mills through his involvement with setting up the first silk *throwing mill* for Thomas Cotchett on the banks of the river Derwent in Derby in about 1704. Throwing is the process by which raw silk is wound from the skein, twisted, doubled and twisted again. The venture failed, apparently through a lack of specialised machinery, but a successful mill was set up on an adjacent site by the Lombe brothers, which marked the beginnings of factory production in England. John Lombe had travelled in Italy, the centre of the silk-throwing industry, and returned to Derby in about 1717 with a number of Italian workmen and pirated details of the most advanced throwing machinery of the time. Thomas Lombe obtained a patent protecting the machines in 1718, and the brothers built a water-powered mill in 1721. The mill was driven by an undershot wheel, and within the building the drive was taken from a pitwheel on the wheelshaft to a shaft that rose vertically through the lower floors of the mill. Drives were taken off by *crown wheels* and pinions to layshafts at each level, a development that was to become standard practice in later textile mills. The origins of this layout are attributed to Sorocold. The lapse of Lombe's patent in 1732 led to an expansion of the industry and more silk mills were built, with Derby, Macclesfield and Stockport becoming centres of production. The Lombe brothers' mill was the prototype for the silk industry nationally, however, and became the model for developments in the cotton industry by Arkwright and Strutt, which took place in the same river valley some fifty years later.

After the dissolution of the monasteries, woollen cloth production increased rapidly, the main centres being Yorkshire, East Anglia, the West Country and parts of the Midlands. Weaving was still a cottage industry and fulling mills continued to be important for finishing broadcloth, as well as for cleaning it before it was dyed. There are many documentary references to fulling mills, and the buildings are sometimes marked on early town maps, often identifiable by the proximity of rack fields with their distinctive *tenter frames* on which fulled and dyed cloth was stretched and dried. Remains of the mills themselves, however, are scanty: only two have been excavated and

Fulling stocks and waterwheel at Higher Mill, Helmshore, Lancashire.

the identification of both sites was reliant on documentary information. At Ardingly, West Sussex, a brick and stone fulling mill powered by an undershot waterwheel had been built in the late seventeenth or early eighteenth century over the site of an earlier water-powered forge. At Hennard Jefford, by the river Wolf in Devon, the remains of a fulling mill of similar age to Ardingly were found during the making of Roadford reservoir. This mill is thought to have been powered by an overshot waterwheel of about 9 feet (2.7 metres) in diameter. At neither site were the remains within the buildings adequate to show how the fulling stocks had been arranged. One reason for the paucity of remains of fulling mills is the continued use of sites for later textile factories, as in the Cotswolds and the north of England, or for other purposes, such as corn milling or papermaking. Often the only clue to the previous use of a mill site for fulling is in the survival of a name such as 'walk' or 'tucking' mill.

In England the production of linen was overshadowed by that of woollen cloth, but the linen industry was important in Scotland and Ireland, where from the early eighteenth century machinery for preparing flax was driven by waterwheels. The flax was first allowed to decompose in order to separate the fibres from the rest of the plant (a process known as *retting*). The fibres were then broken on grooved rollers and beaten with wooden blades fixed to a shaft to remove pieces of straw. By the end of the eighteenth century there were hundreds of these small *scutching mills* in Ulster, including some driven by wind. The finishing processes of *washing*, in which the woven cloth was pounded with water-driven mallets, and *beetling*, using a series of vertical timber stamps to produce a fine sheen, were also water-powered and almost certainly derived from fulling mills. Linen was the most widely used cloth made from plants until the mechanisation of cotton spinning in the late eighteenth century, when the demand for linen declined as cotton was cheaper to produce, lighter to wear and easier to launder.

Papermaking

One industry that made use of established fulling-mill sites was papermaking, for fulling stocks could be readily adapted for pounding rags to make pulp. Papermaking using water power was first established in the 1490s by John Tate at Sele Mill, on the river Beane near Hertford, encouraged by a visit and financial support from Henry VII. Tate's mill fell idle within a few years, however, probably because of competition from abroad. Other paper mills were set up in England in the mid sixteenth century, such as at Bemerton, near Salisbury, in about 1554, the first in Wiltshire and claimed by John Aubrey to be the second paper mill established in England, but it was not until the end of that century that the industry began to flourish. In 1588 the first paper mill to be permanently established near London, at Dartford, Kent, was set up by John Spilman. Spilman, a German, was a jeweller to Elizabeth I and had been granted a licence in 1595 to set up mills on barges on the river Thames. He seems to have been in an advantageous position as far as establishing mills was concerned, for he was granted a monopoly for the manufacture of white writing paper and was subsequently knighted when James I visited his paper mill in 1605. About a hundred paper mills were established in England before the end of the seventeenth century and initially their sites were widely scattered, the main requirements being a good, clean water supply, particularly for making white paper, and access to raw materials. Fine linen and cotton rags were used for white writing and printing papers, and poorer-quality rags, netting and canvas for the more common brown and blue papers that were used for wrapping and packing. Many of the early mills were small-scale enterprises and it was not uncommon for paper mills and corn or fulling mills to be built adjacent to each other or under the same roof. At Sutton Courtenay, Oxfordshire, there were a paper mill and four corn mills under one roof in 1693. The later industry became more concentrated in certain areas, often using former fulling-mill sites, as along the Culm valley in Devon and the Loose stream, near Maidstone, Kent, where paper mills are still at work.

Land drainage and the introduction of the smock mill

In the Netherlands wind power had been used to pump water from the land as a vital part of reclamation schemes from the early fifteenth century, and the introduction of drainage mills appears to be the result of schemes to drain the fens of eastern England from the late sixteenth century onwards. Early references to 'engines' do not specify the power source, and both human muscle and animal power were used before wind was adopted. There were several early speculators who took out licences for drainage engines during the second half of the sixteenth century, including Peter Morrice, who built the water engine at London Bridge. In 1592 one Guillame Mostart petitioned Lord Burghley for the right to drain certain fens and described a new mill that he had constructed to drain the Lincolnshire estates of John Hunt, claimed to be of a kind never before seen in England. A map of 1605 shows Hunt's engine

Walter Blith's drainage engine, from 'The English Improver Improved', 1652.

at Penny Hill, Holbeach, and although the position of the engine is marked by a post-mill symbol, a convention often used by cartographers, it may indicate the site of one of the first *smock mills* to be built in England. Mostart asked for a twenty-one-year monopoly to prevent others from imitating or counterfeiting in whole or part the octagonal, square or other shape of the engines that were being constructed and erected by his direction. The reference to an octagonal shape is significant, for the smock mill is basically a timber-framed tower mill, generally on an octagonal plan. Its origins appear to lie in drainage work, for it could be built with advantage on unstable ground that would not support a masonry tower and it also avoided the need for setting up the heavy trestle of a post mill, including boring through the length of the post so that a drive could be taken down to ground level to work a *scoop wheel*. The greater height and stability of a timber tower also made the smock form a useful one for a drainage engine.

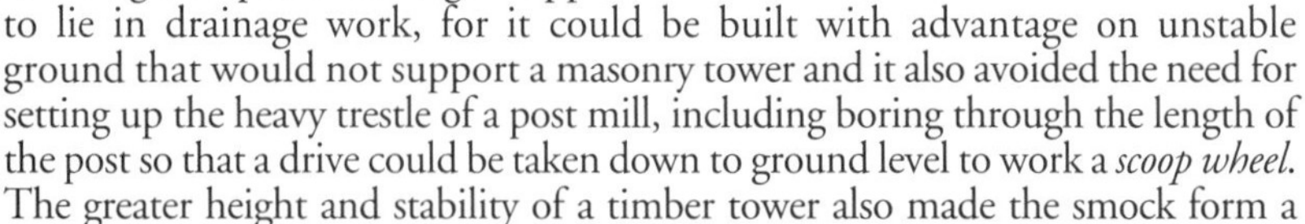

Walter Blith's *The English Improver Improved* of 1652, which deals with aspects of agricultural improvement such as fen drainage, includes the use of windmills for raising water and gives an illustration of a drainage mill. This is extremely crude and does not indicate how the sails could be turned to face the wind, for example, but it does show a timber-framed tower with a scoop wheel for raising water, driven by two pairs of gears from the sails. Scoop wheels, which are constructed in a similar manner to undershot waterwheels, were widely used in fen

Drainage mill at Soham Mere, Cambridgeshire, in the 1950s. Note the circular casing around the scoop wheel and the braced tailpole for winding the cap and sails.

drainage works, and although the height of lift of the early mills was relatively small, about 5–6 feet (1.5–1.8 metres) maximum, this was not a significant problem in England, where depths were not as great as in the Netherlands. Conditions were different in England, however, in that fenland mills usually lifted water into rivers that were liable to flood or where the water level would be regularly affected by tides. Double lifts were used where a greater height was required, a large mill being built near the main river or drainage channel, fed by a smaller mill some distance behind it. One notable difference between English and Dutch smock mills is in the cladding material, Dutch mills being thatched with reed, whereas English mills are clad with timber boards, often placed vertically on drainage mills. The name 'smock mill' is said to originate from the similarity in appearance between the body of the mill and a countryman's smock; if so, it seems likely to be derived from the white under-smock worn by women in the seventeenth century, rather than the decorated, Sunday-best smock of the Victorian farm worker.

Oil milling

Walter Blith also considered that reclaimed land that might be occasionally flooded was good for the cultivation of oilseed-bearing plants, such as cole and rape, and also flax, for linseed. Oil was an important commodity, used for lighting, lubrication and in the preparation of wool for spinning. References to oil mills in the area around Hull, East Yorkshire, occur from the early fourteenth century, and a wind oil mill is mentioned in the will of John Henryson of Hull in 1525. Sir William Dugdale, preparing notes for a history of fen draining during an itinerary of 1657, recorded '4 windmills, used for bruising of Rapeseed, and making oyle thereof; which Rapeseed flourisheth

Water-powered edge runner stones in an oil mill, from Andrew Gray's 'The Experienced Millwright', 1804.

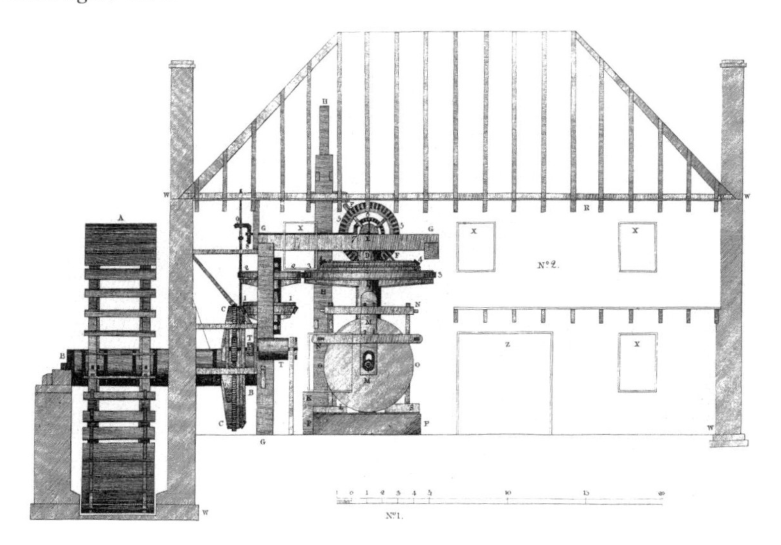

much in this rich fenny country, now that it is thus drayned.' The first references to oilseed-crushing mills in Gainsborough, Lincolnshire, a later centre of the industry, occur at about the same time. The basic equipment of oil mills comprised a set of *edge runner stones*, for crushing or bruising the seed, and stamps, for crushing and pressing to express the oil after it had been bruised and heated. Edge runners consist of two vertical, circular stones mounted on a vertical shaft that roll around a circular bedstone, on which the material to be crushed is spread in batches. Wind power was used for crushing rape seed for oil at Devizes, Wiltshire, by 1716, and water-powered oil mills were introduced into Scotland at about the same time.

Powder mills

Gunpowder, a mixture of saltpetre, charcoal and sulphur, was made in England from the fourteenth century, using some imported ingredients, and ready-made powder was also brought from abroad. Hostile relations between England and Spain during the reign of Elizabeth I, however, gave rise to the need to obtain all the ingredients at home. Water-driven powder mills are known in Britain from the sixteenth century, such as on the Thames at Rotherhithe in 1543, and the majority of gunpowder mills were situated near London, where the Board of Ordnance was based. There were major centres of production in the Lea valley, Essex, and around Faversham, Kent, and the new mills usually took over established water-power sites. At Three Mills, Bromley by Bow, on the river Lea (originally in West Ham, Essex), there is brief mention of a gunpowder mill in 1588, the first in Essex and one of the earliest in England. It was a short-lived venture, apparently making use of an established oil mill on part of an ancient and complex site of tide mills used mainly for grain milling. Gunpowder production began at Waltham Abbey,

Gunpowder mills at Cherrybrook, Dartmoor, Devon. The central chamber contained a waterwheel that drove incorporating mills on both sides.

also on the river Lea, in the mid 1660s, taking over the site of a late medieval fulling mill. The mills were in private hands until the late eighteenth century, when they were purchased by the Crown and subsequently became one of the most important powder works in Europe.

Saltpetre was supplied as fine crystals, but the charcoal and sulphur needed to be crushed, which was done by water-powered edge runners. The powdered ingredients were sieved and then mixed to produce a green charge. Mixing was a risky process, initially done using pestles and mortars, at first hand-driven and later water-powered. *Incorporating mills* using edge runner stones were probably first introduced at Sir Polycarpus Wharton's mills at Wooburn, Buckinghamshire, in the 1680s, and pestle mills were subsequently made illegal on grounds of safety. The edge runner stones were originally driven from above, but the gearing was later moved to below to reduce the risk of accidents. The danger of explosion was always present, and great care was taken to avoid bringing dust and grit into contact with the gunpowder, also guarding against sparks from iron objects and heat caused by friction. The buildings were usually spaced widely apart, and incorporating mills were flimsy structures, often sited amongst trees that would help to absorb the blast in the event of an explosion.

Metalworking

The tasks of extracting and processing metals are both arduous and time-consuming but are two of the fundamental activities of human society, the products of metalworking covering a wide range of uses, both decorative and functional. Water was initially important for washing ore rather than for power, but during the fifteenth century waterwheels were used to drive stamps for breaking up ore, bellows for smelting it and hammers for preparing metal for use. One problem when looking for signs of early mineral extraction and working sites is that the field evidence has often been completely obliterated by more recent intensive working, which leads to problems of interpretation and therefore a reliance on documentary sources and contemporary descriptions. One of the most significant works is Agricola's *De Re Metallica* of 1556, the first book on mining to be based on field research and observation. It is illustrated with

Water-powered ore-stamping mill from Agricola's 'De Re Metallica', 1556. The waterwheel lifts the stamps, D, by cams, H, which project from the wheelshaft. Batches of ore are shovelled under the stamp heads, E.

Blowing house for smelting tin at Week Ford, Dartmoor, Devon. The wheelpit is to the left of the ruined building.

Mortar stone for tin stamps at Week Ford, showing the recesses made by the action of three vertical stampers.

a great number of lively woodcuts depicting German mining and metalworking practice at the end of the Middle Ages, including animal-, wind- and water-powered machines. Considering the influence of German miners on English technology during the later sixteenth century, Agricola's work is an important source of contemporary information.

Tin. The extraction and processing of tin constituted an important industry in Devon and Cornwall, perhaps from prehistoric times, although archaeological evidence is lacking. Tin was a vital ingredient of bronze, when alloyed with copper, and, later, of pewter, when alloyed with other metals such as lead. The early industry relied on alluvial rather than mined deposits, and by the early fifteenth century the ore was smelted in *blowing mills*, which comprised a small building, often built of granite boulders, with a furnace blown by bellows powered by a waterwheel. A 'Blouynghous' is recorded at Lostwithiel, Cornwall, in the early 1400s, and at Dartmeet on Dartmoor a 'blowyng myll and knakkyng myll' were referred to as recently built in 1514. The Dartmoor industry reached a peak in about 1524, before declining to virtually nothing by 1650. There was a brief renaissance about 1700, but only two blowing mills survived on Dartmoor by 1730 and the industry became insignificant after about 1740.

Eight heads of Cornish stamps, driven by a breastshot waterwheel at Lock Stamps, Ludgvan, Cornwall.

'Knacking', 'knocking' or 'clash' mills are all probably synonymous terms for *stamping mills*, which were necessary for processing lode tin. They are recorded as being in use in Cornwall by 1400 and Devon by 1504. In 1602 Richard Carew in his *Survey of Cornwall* described them as having 'three, and in some places six, great logges of timber, bound at the ends with iron and lifted up and downe by a wheele driven with water'. The stamps were lifted vertically by cams projecting from a waterwheel shaft and fell by their own weight on to the ore, which was placed on a mortar stone, usually a block of granite, beneath them. Originally the process was dry, the ore being stamped in batches that were removed manually after crushing. The finest ore seems to have been further reduced by grinding it in a *crazing mill*, which was similar to a corn mill, with a waterwheel driving a pair of small-diameter horizontal millstones. While many stamping-mill sites have been found on Dartmoor, only three crazing mills are known and there seems to have been a general demise in crazing in Devon by the second half of the sixteenth century, although it continued in Cornwall until the late seventeenth century. Wet stamping, illustrated by Agricola and probably introduced from Europe in the early sixteenth century, was a continuous process in which water was used to carry the crushed ore away into *buddles*, where it was allowed to settle. Archaeological excavations carried out at West and East Colliford, Cornwall, produced evidence of wet stamping and indicated that the West Colliford mill was of late-fifteenth-century origin and worked into the early seventeenth century. In the 1990s a considerable amount of work has been carried out by the Dartmoor Tinworking Research Group, including excavation of the Upper Merrivale tin mill in the Walkham valley on the west side of Dartmoor. At least a hundred and fifty stamp mills are known to have been built in Cornwall before 1700, although only three remains have been located, and over sixty are known on Dartmoor, with remains at thirty-two sites. From the remains of the buildings and watercourses, it appears that tin-mill waterwheels were on average about 9 feet (2.8 metres) in diameter by 1 foot 6 inches to 2 feet (0.46 to 0.6 metre) breast and were probably overshot. At the majority of sites where remains can be found the surviving evidence is limited to stone walls, traces of watercourses and wheelpits and, occasionally, stone bearings, mortar stones from stamps and mould stones for casting tin ingots.

Copper and brass. The influence of European workmen and technology was felt particularly in the copper and brass industries during the second half of the sixteenth century with the formation of the Company or Society of Mines Royal, which received encouragement from the state. German miners were employed in the Coniston area of Cumbria in the last decades of the sixteenth century, reworking copper lodes in Newlands and Keswick, although most of their efforts have been obscured by later mining. Water power was used for both pumping and stamping and there are references from the 1560s and 1570s to stamping mills being built by English carpenters under the direction of their German counterparts. Copper was also mined in Cornwall later in the sixteenth century, where its extraction and smelting are closely comparable to that of tin, with which it was found.

The technology of making brass wire and sheet was brought into England from Germany and southern Flanders, where there was a well-established industry. The Mineral and Battery Company (where 'Battery' refers to battery ware, meaning pans and dishes) was given a monopoly of brass manufacture in 1567, and water-powered forges were set up at Tintern, Monmouthshire, although this early venture was not a success and the forges eventually concentrated on making iron wire and sheet. With the rebirth of the English copper industry towards the end of the sixteenth century, a number of copper and brass mills were established in the Bristol area, using ore from Cornwall and calamine, a source of zinc for making the brass alloy, from the Mendip Hills. A brass works was established in 1702 at Baptist Mills, on the river Frome to the north-east of Bristol, by Abraham Darby, using workmen brought over from the Low Countries. Darby withdrew his interest in the brass mills and subsequently moved to Coalbrookdale, Shropshire, where in 1709 he successfully introduced the use of coke to smelt iron. In 1721 the Bristol Brass Company leased the site of a fulling mill at Saltford, on the river Avon between Bristol and Bath, and established rolling and battery mills, where sheets of copper or brass were worked under water-powered hammers to form utensils. Significant remains of the mills, which produced battery ware until 1908, the last to do so in Britain, still survive.

Lead. Water power was also used in the lead industry, both for pumping from the mines and driving stamps and bellows for working the ore. The industry was important in several areas such as the Derbyshire Pennines, the Yorkshire Dales, the northern Pennines, south-west Scotland and central and north Wales and, to a lesser extent, the Mendip Hills in Somerset and the Lake District. Water-powered smelters were in operation in several districts before the end of the sixteenth century and the first to be built in Scotland was at Wanlockhead in 1682. Both horse and water power were used for pumping and winding. In the Yorkshire Dales waterwheels were usually overshot and water was often brought considerable distances to supply them, along timber *launders* or troughs supported on posts. The wheels were usually set in masonry pits with their shafts set at about ground level, with a crank on one end

The water-powered charcoal blast furnace at Dyfi, Machynlleth, Powys. The large-diameter high breastshot waterwheel dates from the furnace's last use as a sawmill.

to drive the pumps and a winding drum at the other end, which could be put in and out of drive by a simple clutch, its motion being controlled by a band brake. Ore crushers were generally placed in the wheelhouse, with a washing floor for ore arranged to utilise the tail water from the wheel. Silver production was often closely associated with that of lead and in an inventory of the equipment of a lead and silver works in Cardiganshire in 1667 four large waterwheels are listed. In the smelting mill one new large waterwheel worked five pairs of bellows to blow five hearths. The stamping mill had a large waterwheel working three sets of stampers and an 'annexed mill to grind bone ash with a pair of stones, with all things necessary for grinding and sifting bone ash in order to refine'. The refining mill also had a large new waterwheel driving three pairs of bellows, and the red lead mill had one great waterwheel that drove four pairs of stones for grinding red lead, an early example of the development of multiple drives from a single prime mover.

Iron. Iron became an increasingly important commodity from medieval times, being used in building, agriculture, shipping and arms. The early industry was concentrated in the Weald and the Forest of Dean in the south, and in Yorkshire, Weardale and Furness in the north of England. It relied on local sources of ore, of wood to make charcoal for smelting and on the availability of water power. Iron ore was initially converted in a simple furnace called a *bloomery,* which, from the thirteenth century, was blown by water-powered bellows and was capable of producing small amounts of wrought iron direct from the ore as and when needed. The annual output of a bloomery was probably about 20–30 tonnes when in regular operation. The bloomery process persisted in some areas such as the Forest of Dean into the early seventeenth century and in the north-west of England to about 1700 but has left few remains. At Rockley Smithies, South Yorkshire, a water-powered bloomery first built in about 1500 was excavated in 1964–6. The forge had undergone a major rebuilding about a century after it was established, with the construction of a bloom hearth and probably two string hearths, on which products such as scythes, spades, javelins and other functional ironwork were made. Remains of water-powered bellows and fragments of two waterwheels, made largely of oak, were also found. The waterwheels were overshot; one was about 11 feet (3.4 metres) in diameter and had probably replaced a broader breastshot wheel as the width of its wheelpit had been reduced. Power requirements were relatively limited and the waterwheels would have probably developed between 1 and 2 horsepower (1 kW). The forge was demolished in about 1640, perhaps being quarried shortly afterwards for material for a blast furnace built a short distance away in 1652.

The first English blast furnace was established in 1496 at Newbridge, Ashdown Forest, and several early examples have been excavated in the Sussex Weald, notably at Panningridge, Chingley and Maynards Gate. At Panningridge fragments of two waterwheels were found, one of which was overshot and about 10 feet (3.1 metres) in diameter by 15 inches (0.4 metre) wide. Blast furnaces relied on waterwheels for working bellows, and

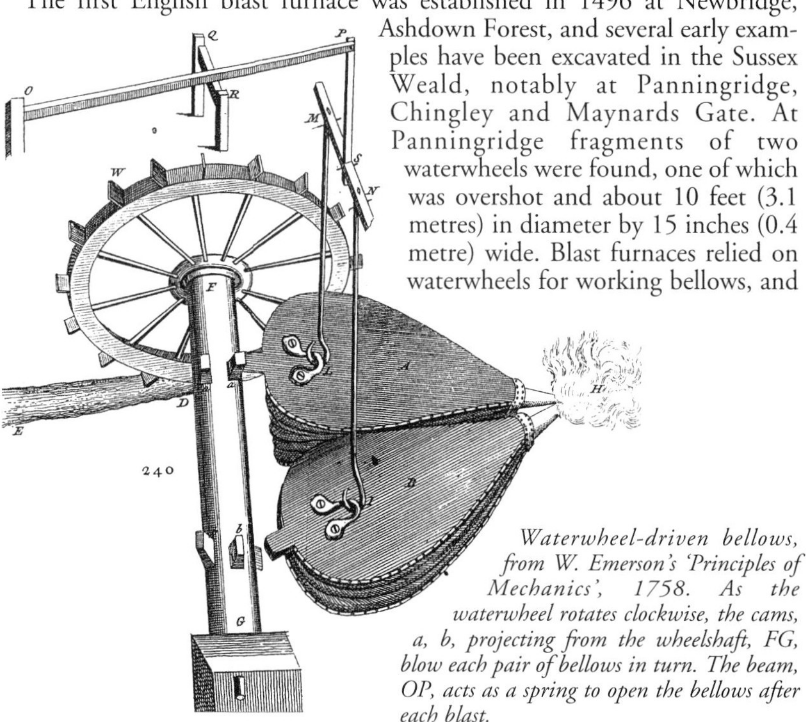

Waterwheel-driven bellows, from W. Emerson's 'Principles of Mechanics', 1758. As the waterwheel rotates clockwise, the cams, a, b, projecting from the wheelshaft, FG, blow each pair of bellows in turn. The beam, OP, acts as a spring to open the bellows after each blast.

Heavy forging hammer at Wortley Top Forge, South Yorkshire. The frame of the waterwheel, before restoration, can be seen to the right. The hammer was tripped by a camwheel located just behind its head.

their sites can sometimes be located by ponds and the earthworks of dams. The bellows were sited alongside the wheelpits and worked by cams projecting from an extended wheelshaft, the remains of which were found at Chingley. Although the early iron industry was concentrated in the south-east, the emphasis changed after the introduction of the blast furnace, and water-powered furnaces were built in Shropshire, Staffordshire, Glamorgan and Monmouthshire in the 1560s. Many early furnaces were built to supply the Crown with arms, and a survey of the Crown Ironworks in the Forest of Dean in 1635 describes the furnaces as having waterwheels of about 23 feet (7 metres) in diameter to drive bellows. The wheels were usually overshot and at one furnace water was brought to the wheel along a raised timber trough some 75 yards (68 metres) long, cut out of tree trunks.

Molten iron was run from the blast furnace into a casting house, which, like the bellows house, was probably a lean-to structure built up against the stone walls of the furnace. The floor of the casting house had channels in which pig iron was cast and also a casting pit, where moulds for objects were put. Cannon were cast hollow in deep vertical pits and then bored out. Only one cannon-boring site has been excavated, at Pippingford, Sussex, where there was no evidence of how the boring mill was powered, although water-driven examples are known from documentary sources. Cast-iron pigs were converted into wrought iron by the 'finery' process. A *finery* was a conversion forge, often utilising a former bloomery site, usually with two or three waterwheels. One drove bellows for a hearth to remelt the pig iron to convert it to wrought iron, using a blast of air to burn out the carbon. A second waterwheel drove a hammer to forge the slag from the bloom, and there was sometimes a third for the 'chafery', a hearth where wrought iron that had cooled during production under a tilt hammer was reheated for further forging. In 1631 Lydbrook forge in the Forest of Dean had two fineries, each with a waterwheel driving bellows, a great hammer worked by a waterwheel, and a chafery wheel. Several fineries have been excavated in the Weald, at Ardingly, Blackwater Green and Chingley, for example. Wortley Top Forge, by the

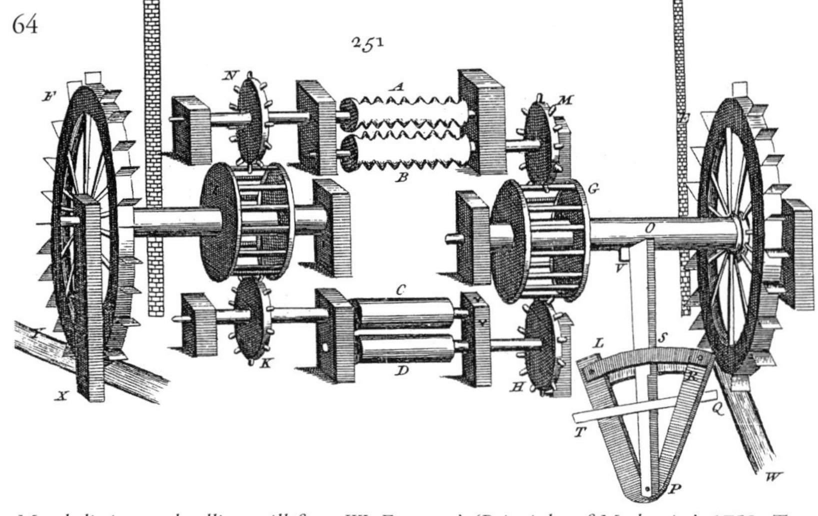

Metal slitting and rolling mill from W. Emerson's 'Principles of Mechanics', 1758. Two undershot waterwheels, on opposite ends of a building, drive a pair of toothed rolls, AB, for slitting bar and a pair of smooth rolls, CD, for making plate or sheet. The waterwheel on the right also works shears, SP, by a cam, V, on its wheelshaft, to cut bar, TQ.

river Don in South Yorkshire, was used as a finery forge in the first half of the eighteenth century and the tilt hammer that survives is a notable example, although many of its original timber components have been replaced in iron.

The use of mineral fuel to replace charcoal, the production and supply of which became increasingly under pressure as more ironworks were established, was tried from the early seventeenth century, but without success. Coke was first successfully used in 1709 at Coalbrookdale, Shropshire, in a furnace leased by Abraham Darby in 1708. Although later rebuilt, the remains of his original furnace attract great attention as one of the most significant relics of the early industrial revolution. Like other blast furnaces of that period, it relied on a waterwheel to work the bellows.

By the end of the sixteenth century water power was also used to drive machinery for rolling and cutting metal. Thin sheets of rolled wrought iron were required for making armour, and iron and brass sheet was also rolled and cut into strips in water-driven slitting mills, the first in England being set up at Dartford, Kent, in 1590. Strips of iron were used for making nails, mainly for building, and horseshoes and could be further drawn out to make wire, important for making carding, a form of fine-toothed comb or brush used to prepare wool for spinning. Although a wide range of metal objects was still made and finished by hand, water power played an increasingly significant part in the production of items such as tools and cutlery. The earliest reference to grinding metal by water power on the river Sheaf in Sheffield, later the centre of the cutlery industry, appears in 1496. The early history of the industry is not well documented but by the sixteenth century the tenants of the grinding wheels often combined industrial and agricultural pursuits in order to provide a living.

Overshot waterwheel with flat rod drive for pumping at Wheal Martyn china clay works, St Austell, Cornwall. The rod extending forward from the crank on the waterwheel shaft is connected to the balance bob, top left.

Coal mining

It is perhaps surprising to consider the historical importance of water and wind power in the coal industry. As with many other extractive industries, coal mining has early origins, and coal was exploited by landowners, including monasteries, in the Middle Ages. Power was needed for winding or haulage of both men and materials and for pumping. Winding was done by man-powered windlasses on small shafts and by horse gins on larger pits. Pumping and drainage became important as pits became deeper, and waterwheels, horse gins and even wind pumps were used before the beginning of the seventeenth century. In Tyne and Wear large-scale mining activity made mechanical drainage increasingly necessary by the late sixteenth century, and there was a *coalmill* at Wollaton on Trent, Nottinghamshire, by 1580. In about 1667 Stephen Primatt wrote that 'In most collieries in the North they make use of chain pumps, and do force the same either by horse wheels, tread wheels, or by water wheels.' Coalmills were waterwheel-driven pumps, using buckets or disc chains to raise water. A chain engine drew water up standing pipes bored out of elm-tree trunks by discs mounted on a continuous chain. A great deal of power was lost in transmission, however, and the depth of lift was restricted to about 15 fathoms (27 metres). As most parts were still made of timber there was also a need for frequent repairs.

After the mid seventeenth century, the maximum effect of water power was developed using a sophisticated treble shaft system, for example at Ravensworth near Durham, which had three waterwheels measuring 24 feet (7.3 metres) in diameter working in series, and by the late seventeenth or early eighteenth century Lumley Colliery, on the river Wear, had nine waterwheels. The inefficient rotary drive of the chain engine was, however, eventually replaced by rod drives to banks of reciprocating pumps, which appeared by about 1700. These engines were known as bob gins, from the reciprocating motion of their pump beams, and enabled water to be lifted from more than 40 fathoms (73 metres). They were in common use on the Durham coalfields until the mid eighteenth century. The rod engine or 'stangenkunst' was known in Europe from at least 1540, using *flat rods* connected to a crank on the end of the waterwheel shaft.

Flat rods are so called because of their working position rather than their shape and were made originally of timber and later of wrought iron. They were held in tension by bobs placed near the waterwheel that partially balanced the weight of the vertical pump rods. They were in general use by the 1570s and were probably first introduced into England by German miners working the copper mines of Cumbria at about that time.

On the Somerset coalfield, waterwheel-driven pumps are first recorded by 1610, and in 1700 a waterwheel of 8 feet (2.4 metres) in diameter by 3 feet (0.9 metre) wide, driving two cranks, drew water from about 17 fathoms (31 metres) at Mr Brewer's works at Paulton. In the 1680s a post mill was built to drain coal workings at Whitehaven, Cumbria, by John Satterthwaite, a carpenter, who had been sent to the Newcastle area to see the coalmills there. The drive for the pumps was taken through a hole bored through the centre of the post of the mill, a rare example in England of a *hollow post mill*, a type known in the Netherlands as the 'wipmolen', where it was originally developed for land drainage. Satterthwaite's windmill was probably not a success as it was superseded by a steam engine in 1715. An agreement of 1700, between a millwright, a gentleman, a yeoman and a collier, resulted in the building of a wind engine to draw water by pumps from a depth of 12 fathoms (22 metres) at South Park, Wraxall, Somerset. The engine cost about £41, and £2 a year were allowed for its maintenance over the next seven years. In Scotland there were proposals to build wind gins to drain coal workings during the early eighteenth century, and John Young, a millwright of Montrose, was sent to the Netherlands to inspect machinery there. Windmills were undoubtedly powerful enough for drainage work but during calm periods the mines flooded and workmen were made idle. Both water and wind power were generally superseded for pumping by the introduction of Thomas Newcomen's steam engine after 1712. The replacement of existing coalmills with steam pumps was not immediate, however, for the mills were relatively cheap and perfectly adequate for the work they undertook, although their maintenance costs were apparently high.

Until the middle of the eighteenth century the majority of industrial processes in Britain relied on water rather than wind power. Of the number of other trades and industries that adopted both water and wind power to work machinery in the seventeenth century, little physical evidence survives. Timber sawing, the processing of imported raw materials for dyestuffs and snuff grinding will be considered later, being essentially connected with the period known as the industrial revolution.

THE AGE OF INDUSTRIALISATION

The period of great technological development in eighteenth-century England commonly known as the industrial revolution was carried forward using the power of water and, to a lesser extent, wind. By the time the steam engine became an economic alternative early in the nineteenth century most of the technical improvements to water- and wind-powered machinery, prime movers and millwork had already been made. Before 1800 an average steam engine was capable of producing no more than 18 horsepower (13.5 kW), which was within the capacity of a waterwheel or a well-found windmill in a stiff breeze. An inevitable drawback of both water and wind power, however, was the restriction of siting, particularly in growing urban areas. Water was sometimes conveyed considerable distances to supply industrial sites, and the hydraulic systems of reservoirs, ponds and leats, although often long disused, have left impressive remains in the landscape. In mining areas such as Cornwall a complex artificial drainage pattern became superimposed on the natural one, the full extent of which will never be known as it was in a constant state of change and many workings were short-lived.

The state of European millwrighting at the beginning of the eighteenth century is well illustrated in a number of books about mills. These include Jacob Leupold's *Theatrum Machinarum Molarium* (Leipzig, 1735), the books of the Swedish millwright Linperch and those of his Dutch contemporaries van Natrus, Polly, van Vuuren and van Zyl (published in Amsterdam between 1727 and 1741) and Belidor's *Architecture Hydraulique* (Paris, 1737–9). No comparable works exist in English but the influence of European practice was evident in many trades and processes and it is probable that the technology of British mills was similar to that on the European mainland. Timber was still the main material used in the construction of machinery, and the design of waterwheels and windmill sails relied heavily on empirical practice. Dutch millwrights advised on the design of wind-powered oil mills in Hull in the 1740s and were probably responsible for the introduction of the tall, brick tower mill into East Yorkshire, from where it subsequently spread throughout eastern England. The importance of European technology was soon overshadowed, however, by developments in Britain. Several of the early engineers, whose names have become familiar for their achievements in the wider fields of civil and mechanical engineering, made significant contributions to the use of water and wind power and to improving the efficiency of waterwheels and windmill sails. Their influence was also felt in traditional millwrighting, particularly with the introduction of cast iron into machinery.

John Smeaton (1724–92) was the leading engineer working in the field of water and wind power in Britain during the second half of the eighteenth century. He began his career making scientific instruments but had a strong interest in engineering and in 1752–3 undertook a series of experiments into the power of waterwheels and windmill sails, using working models. After a

John Smeaton, FRS, 1724–92.

period when he put some of his findings into practice, he presented his experiments to the Royal Society in 1759, for which he was awarded their Copley Medal. He wrote a paper entitled 'An Experimental Enquiry concerning the Natural Powers of Water and Wind to turn Mills and other Machines depending on a Circular Motion', which was later published in book form and quoted in many nineteenth-century works on hydraulics and pneumatics. The paper did a great deal to establish his reputation as a consulting engineer. It is evident from his diary of a *Journey to the Low Countries* (1755), in which he refers several times to visiting mills which 'differed in nothing from the description in the Dutch mill books', that he was familiar with contemporary Dutch millwrighting and this influence can be seen in some of his designs, although he was able to combine established practice with innovation.

Smeaton's designs, which are preserved by the Royal Society, and the *Reports* of his work as a civil engineer, published posthumously in 1812, provide a valuable record of the development of water and wind power in the second half of the eighteenth century. A lesser-known but significant book is *The Experienced Millwright* by Andrew Gray, printed in Edinburgh in 1804. Little is known about Gray, described in his book's preface as 'a practical Mechanic [who] has been for at least forty years employed in erecting different kinds of Machinery', but his book stands alone in setting out both the principles of mechanics and directions for building mills. It also provides a series of fine engravings showing the layouts of various types of water-, wind- and animal-powered machinery. His examples are all apparently drawn from central-eastern Scotland, but they still provide a valuable visual record of the state of traditional millwrighting in the second half of the eighteenth century.

The introduction of cast iron

The early use of cast iron for gearing and shafting in millwork is generally credited to John Smeaton. He introduced iron gears at the Carron Ironworks, Stirling, Scotland, in 1754 and his designs of 1754–5 for a wind oil mill at Wakefield, West Yorkshire, show a cast-iron windshaft, although the rest of the gearing and shafting is of timber and rather conventional in appearance. One reason for the introduction of cast-iron shafting was an increasing shortage of timber of suitable size, as well as the longer-lasting potential of iron when in contact with water, but there were problems with the new technology and several early cast-iron shafts failed owing to flaws in the castings. Smeaton had some of his shafts cast at the Carron Ironworks. The cost of transporting them must have been a further major consideration. The use of iron for the machinery of transmission was initially slow to develop, partly because of the technical difficulty of casting large gears, but iron pinions were used by John Rennie at Bonnington Mills near Edinburgh in 1780 and he also designed the millwork for the steam-powered Albion Mills in London in 1784. According to the industrial biographer Samuel Smiles, 'The whole of the wheels and shafts of the Albion Mills were of iron, with the exception occasionally of the cogs, which were of hard wood, working into other cogs of cast iron; but where the pinions were very small they were made of wrought iron.' This arrangement, where gears with inserted hardwood cogs mesh with those with integrally cast iron teeth, is found in many surviving mills and proved to be a good working compromise. Timber cogs could be accurately pitched and trimmed, making for smoother and quieter running, particularly in gearing where the pitch (the spacing between the teeth) is coarse, and they are also easier to replace when worn or damaged.

Late-eighteenth-century cast-iron pinions driven off a timber crown wheel in White Mill, Shapwick, Dorset.

Millwrights were quick to appreciate the benefits of iron, particularly for those components that needed to withstand constant wetting or weathering. In 1785 a millwright in the Halifax area agreed to build a predominantly timber overshot waterwheel on an oak shaft, but the flanges or *naves* that connected the arms to the shaft were to be of cast iron and 'fitted to the axle tree without mortise', thus alleviating the problem of water penetrating into the mortises and eventually rotting the shaft. At Wheatley tower mill, Oxfordshire, the *poll end* or *canister*, by which the sail stocks are connected to the end of the timber windshaft, is of iron and carries the date 1784. It was cast at the Eagle Foundry, which was established in Oxford in about 1760. One of the earliest waterwheels to be made completely of iron was 'Aeolus', a 50 foot (15.2 metre) diameter wheel built in 1800 by Watkin George at the Cyfarthfa Ironworks, South Wales, for blowing three blast furnaces. Iron was used increasingly for waterwheel construction from that time. The Navigation Mill at Warwick had a 'great wheel' of 24 feet (7.3 metres) in diameter made by Mr Roberts of Warwick in 1805, described as being of cast iron and of 'excellent construction'. By 1820 iron castings were widely available from small foundries that were opening up throughout the countryside. Many millwrights continued to work in timber, however, and the survival of wooden gearing in some areas is probably due to the persistence of local millwrighting traditions as well as the cost of castings and their transport.

Cast iron became increasingly important as a material for steam engines, bridges and also the frames of mill buildings in the last quarter of the eighteenth century, and its production and use were closely connected with those innovators who were prepared to use it. Thomas Telford (1757–1834), an engineer noted for his bridges in particular, wrote a treatise *On Mills* for the Board of Agriculture in 1796–8, when he was based at Shrewsbury. It was apparently intended for publication and drew together a great deal of information, including published sources such as Smeaton's work. Telford's manuscript includes some detailed drawings of both watermills and windmills built in the 1780s by William Hazledine of Shrewsbury, 'a very ingenious practical Millwright'. Hazledine was also an ironfounder, who cast several of Telford's bridges and supplied the beams and columns for Beynon's Flax Mill at Castle Foregate, Shrewsbury, which was built in 1796 to the designs of Charles Bage and has the distinction of being the first iron-framed mill building to use both cast-iron beams and columns. The drawings of the corn mills built by Hazledine show the machinery to have been still largely made of timber, although Telford refers to the 'practice now begun of making water wheels of Iron'. He also gives the cost of a three-storey watermill with two pairs of millstones built in the Shropshire–Cheshire area as about £350 and that of a comparable four-storey brick tower mill as about £500.

The development of industrial waterwheels

In the mid eighteenth century waterwheels were built predominantly of timber, with fastenings and other iron fittings made by blacksmiths. An estimate for building a corn mill at Tamworth, Staffordshire, in 1760 refers to there being 'a great deal more work in an Over Shot Mill Wheel, than there is in an Under Shot', the extra work adding about six per cent to a total building cost of £104. Overshot wheels were more efficient, particularly where there was a limited but constant supply of water, making use of its weight rather than simply relying on flow. However, they needed to be built more strongly and required more maintenance as they worked wet and had to support the weight of the water in their buckets. Some of John Smeaton's work on waterwheels was concerned with obtaining greater efficiency for wheels driven by low falls, as only a portion of the power of any fall could be obtained using an undershot wheel. He substituted breastshot wheels where possible, so making use of both the impulse and the weight of the water. One feature of his designs is the use of close-fitting masonry breastwork, so that water was held on the wheel to perform as much work as possible. This feature can still be found in a developed form in the *start and awe wheels* still found in parts of Scotland, notably Aberdeenshire. The starts project from the rings of

Smeaton's design for waterwheels for the Carron Ironworks, 1770. Note the close-fitting masonry breastwork and the use of cast iron for the rings that carry the floats.

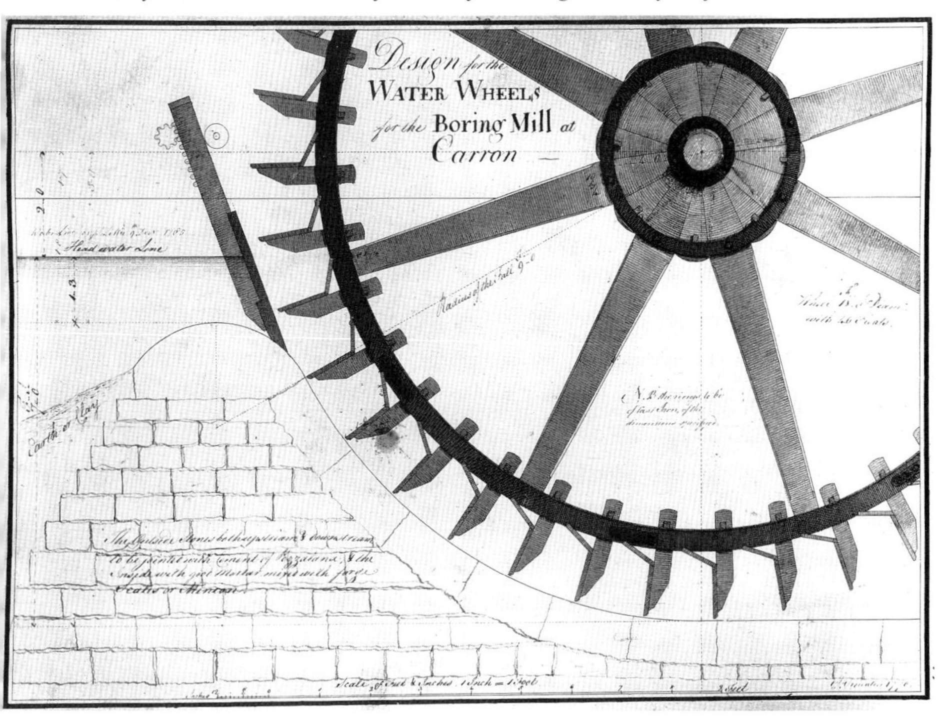

Timber race for water feed to a start and awe waterwheel at the wood-turning mill, Finzean, Aberdeen. This arrangement uses a head of water suitable for an overshot or high breastshot wheel to give a higher rotational speed.

the waterwheel to support the awes, or floats. In principle this type of wheel is undershot, but the water is fed along an inclined, shaped trough from a level that would usually be more conventionally used for an overshot or breastshot wheel. Smeaton illustrates an example at Aberdeen in his *Reports,* and there are good illustrations in Andrew Gray's *The Experienced Millwright*. Smeaton also took considerable care over the design of penstocks, the shuttles or sluices that let water on to wheels, and several of his designs are for *pitchback wheels*, an accurate definition of which is that water is laid on at the top of the waterwheel, in the manner of an overshot wheel, but the wheel turns in the same direction as the flow, as with a breastshot wheel, and is therefore less likely to be impeded by tail water.

Smeaton used cast iron for the rings of waterwheels at the Carron Ironworks in 1770, 'in order that they may act as loaded flies [fly wheels]', and in 1780 proposed the use of iron plates to form the buckets of a waterwheel he designed for Carshalton Mill, Surrey, 'which compared with those shown of Wood will allow the Bucketts to be more capacious & a better Entry for the Water'. His waterwheel designs showed the way forward, both in terms of the use of iron and in his approach to improving efficiency, but, although he worked on a considerable number of mills, little of his work survives. At Canterbury, Kent, two wheel races and a cast-iron wheelshaft on the site of the former City Mill, which was burnt down in 1933, are important remains,

City Mill, Canterbury, Kent. Designed by Smeaton, this fine mill burnt down in 1933. Two wheelpits and a cast-iron shaft still survive on site.

for the mill was built from plans made by him in 1792, the year of his death.

John Rennie (1761–1821), a talented engineer who, like Smeaton and Telford, is now usually associated with civil engineering works rather than water and wind power, spent his early years working for Andrew Meikle, a practical millwright who was a tenant of Rennie's father and had his workshop at Houston Mill, East Linton, East Lothian. In 1783 Rennie invented the *sliding hatch*, a form of waterwheel penstock in which the sluice was lowered rather than raised, so that water flowed over the top. By lowering the hatch as the water level in the millpond or race fell, the maximum fall could be utilised in varying conditions. The idea was apparently suggested by an observation he made as a child when he saw water spilling over the top of a closed sluice gate at Houston Mill. In 1784 Rennie started work for Boulton & Watt, his first job being to construct a new waterwheel for Matthew Boulton's Soho factory in Birmingham, and shortly afterwards he designed the millwork for the steam-powered Albion Mills in London, which started work in 1786. When Smeaton visited the Albion Mills he pronounced them to be 'the most complete, in their arrangement and execution, which had yet been erected in any country'. Rennie also designed two water-powered canal pumping engines, both of which survive and provide an interesting contrast in the use of timber and ironwork. The Melingriffith Pump, on the Glamorganshire Canal in Cardiff, was built by William Jessop in 1807 and originally had a massive

John Rennie's Claverton pumping station on the river Avon, near Bath, Somerset, showing the breast of the wide waterwheel and the control gear for the sliding hatches.

oak wheelshaft with elegant cast-iron naves and a crank from which the pumps were driven. The frame supporting the pump rods was also of timber. At Claverton, near Bath, Somerset, Rennie built a water-driven beam pump on the site of a former corn mill to raise water 48 feet (14.6 metres) from the river Avon to the Kennet & Avon Canal. The engine was completed in 1813 and is predominantly of iron construction, originally with a low breastshot wheel measuring 17 feet 8 inches (5.4 metres) in diameter and 24 feet (7.3 metres) wide. Although mounted on a massive, square, cast-iron shaft, the width of the wheel caused continuous problems due to the shaft bowing, and in 1858 a central bearing was fitted. Water was let on to the wheel through sliding hatches and, running at a speed of about 5 rpm, it was capable of producing 24 horsepower (18 kW).

During the eighteenth century the quest for power led to the construction of bigger waterwheels, particularly for industrial uses such as mining and textiles, until the limits of timber construction were reached. Smeaton reported

A 35 foot (10.7 metre) diameter wheel at Glynn Valley china clay pit, Cardinham, Cornwall, in about 1870. Note the two water feeds and the apron to prevent water being blown off the top of the wheel in high winds. The man has one hand on the flat rods where they are connected to the crank. At the other end of the wheelshaft a winding drum is visible.

that 'the largest I ever saw, was in Cornwall, and forty-eight feet [14.6 metres] diameter; the largest I ever constructed is forty-seven feet [14.3 metres] . . . . for, when for the sake of making the best use of a small quantity of water, it would require the Water-wheels to exceed forty feet [12.2 metres] in diameter, or height, then they become so heavy, unwieldy and expensive, as on those accounts to be ineligible; and they rarely exceed fifty feet [15.2 metres]'. Large timber-built wheels were slow moving and heavily stressed, particularly if they were overshot, for the constant wetting and drying caused problems of balance and accelerated decay. By the end of the eighteenth century it had become practice in the Cornish mines to replace several small waterwheels with a single large one, usually overshot, the optimum diameter being about 38 feet (11.5 metres). Running at a speed of about 4 rpm, these large wheels were well suited to drive pumps for lifting the water out of mines. The pumps were driven by flat rods from a crank on the end of the wheelshaft and, although cast-iron wheelshafts and cranks were in use from about 1800, all-iron wheels were rare before the middle of the nineteenth century and oak was still preferred for arms and shrouds.

In 1803 John Taylor (1779–1863), the 'patriarch of British mining', started work on driving a canal tunnel through Morwell Down, near Tavistock, Devon, to link the rivers Tavy and Tamar. The tunnel was 1 1/2 miles (2.4 km) long, and a 40 foot (12.2 metre) diameter waterwheel was used to pump out water during its construction. Ventilation also proved to be a major problem, so Taylor built three waterwheel-driven pumps, reversing the action of the pumps so that they sucked out the stale air. The tunnel was completed by 1810 and, because the canal finished high above the quay at Morwellham on the river Tamar, an inclined plane was built with a waterwheel driving winding machinery to draw trucks up nearly 240 feet (73 metres). Water power was used elsewhere for lifting purposes, such as the Bude Canal in north Cornwall, where from the 1820s five inclined planes were worked by waterwheels.

Waterwheels were also used to power hoisting machinery in canal warehouses. James Brindley (1716–72) was a millwright of considerable ability who became an engineer

Early-nineteenth-century breastshot waterwheel with the drive taken from a ring gear at Ashton-under-Lyne, near Manchester. The wheel worked hoisting machinery in an adjacent canal warehouse.

Overshot waterwheel with ring gear drive, the second generation wheel at Arkwright's cotton mill, Cromford, Derbyshire.

particularly remembered for his work on hydraulic systems and canals. He built an underground waterwheel on the Bridgewater Canal at Castlefield, Manchester, in the 1760s for hoisting coal up to street level. On the Rochdale Canal at Dale Street, Manchester, a later installation including the remains of a high breastshot wheel was discovered in 1983. The wheel, which drove hoists in two warehouses, is thought to have been designed by Thomas Hewes and was installed in about 1824.

Much of the innovation that took place in the design of waterwheels in the 1790s was due to the need to build more efficient and powerful prime movers for use in the rapidly expanding textile industry. The growth of the cotton industry was particularly spectacular in Lancashire and Yorkshire, where the damp Pennine air helped the cotton fibres to cling together during spinning, the land was cheap and water power was available. All the main processes of spinning and weaving were mechanised during the second half of the eighteenth century, but there was considerable social unrest in the form of machine-breaking riots as handworkers feared that they would be made redundant by the new machinery. Richard Arkwright, the inventor of the water frame for mechanical cotton spinning, moved away from Lancashire in 1768 to avoid the machine breakers. His first spinning mill, which was powered by horses, was in Nottingham, in an area that was already industrialised but where the labour force was more docile. In 1771 he built a second mill at Cromford, Derbyshire, using water from the Cromford Sough to drive an overshot waterwheel. The original wheel, which was fed by an overhead timber launder, was replaced in 1821. Arkwright and his partner Jedediah Strutt were the real initiators of the factory system in cotton spinning, and their development of textile-mill sites in the Derwent valley probably owed much to the work of George Sorocold and the Lombe silk mill at Derby some fifty years earlier. By 1788 there were 122 spinning mills in England and Scotland. After the partnership with Arkwright ended in 1781, the Strutt family concentrated on developing mills in Belper, Derbyshire, and in the first decade of the nineteenth century William, Jedediah's son, made a significant contribution to contemporary technology, in the

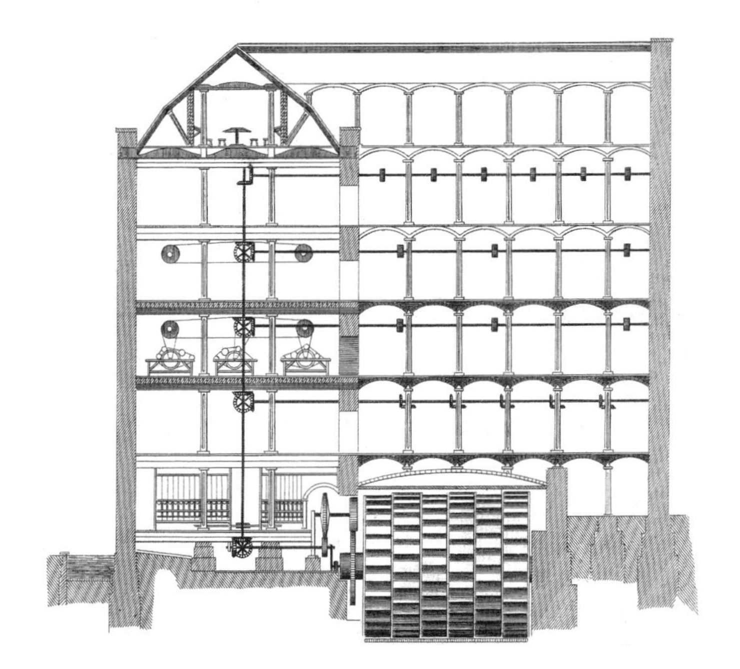

Stylised layout of a cotton mill, based on Strutt's North Mill, Belper, Derbyshire. The waterwheel has staggered buckets to reduce shock as water enters and thus to make for smoother running. Drives are taken from bevel gears to layshafts at each floor level.

development of both a fireproof, iron-framed construction for mill buildings and a new form of an all-iron waterwheel.

Probably the widest wheels ever built of timber were those installed in Strutt's West Mill at Belper in the 1790s. Contemporary descriptions are slightly conflicting, but there appear to have been two breastshot wheels, one of which was 40 feet (12.2 metres) wide by a little over 12 feet (3.7 metres) in diameter. As timber was not available in suitable lengths, the shaft of the wheel was formed hollow from a great number of pieces of timber hooped like a cask or barrel, with floats radiating from it. These massive wheels were relatively short-lived, being replaced within nine years by narrower ones of greater diameter and of a new, relatively lightweight construction in which circular wrought-iron arms and cross braces, held in tension, supported the shrouds. Gear rings were attached to the shrouds, from which drives were taken by pinions and layshafts. The wheelshafts were made of cast iron and cruciform in cross-section, being designed in the form of a rotating beam, rather than simply being based on empirical practice. By taking the drive from a large-diameter ring gear and small pinion, a greater rotational speed was achieved without a large train of gears, which was of great benefit for driving textile machinery, and the size of the wheelshaft could be reduced as it was no longer needed to transmit the drive. Two of these *suspension wheels* were installed at Belper between 1804 and 1810 and, while the idea appears to have been Strutt's own, the practical work and early development were carried out by the millwright and textile-mill engineer Thomas Hewes (1768–1832).

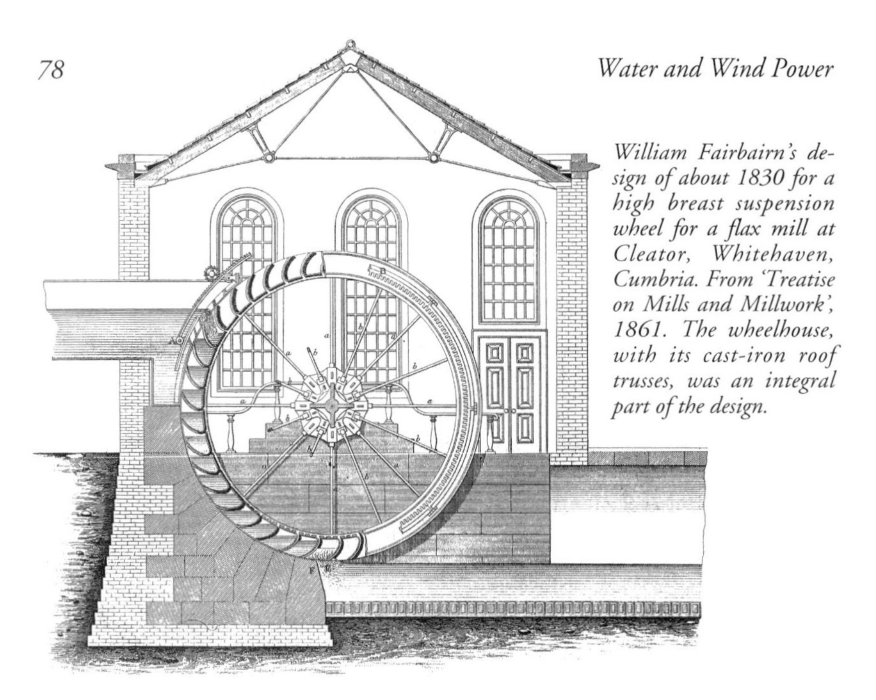

William Fairbairn's design of about 1830 for a high breast suspension wheel for a flax mill at Cleator, Whitehaven, Cumbria. From 'Treatise on Mills and Millwork', 1861. The wheelhouse, with its cast-iron roof trusses, was an integral part of the design.

Hewes was working in Belfast in 1790, probably setting up a cotton-spinning mill, but moved to Manchester, the heart of the textile industry boom, in 1792. He worked with Strutt in the first decade of the nineteenth century and brought together a number of different features that seem to have been used in isolation before the development of the suspension wheel. He was also instrumental in developing the *governor*, already patented for use in windmills and on steam engines, to control penstocks. The application of governors to regulate the speed of textile-mill waterwheels was quickly appreciated for, in the words of Robertson Buchanan in his *Practical Essays on Mill Work* (1823), 'In the case of a cotton-mill . . . which is calculated to move the spindles at a certain rate, if from any cause the velocity is much increased, a loss of work immediately takes place, and an increase of waste from the breaking of threads.' In 1821 Hewes was joined in partnership by Henry Wren, a former employee, and by 1824 they were employing about a hundred and fifty men, forty on heavy millwork and over a hundred on machines, heavy gearing and fire-proof mills. Their Manchester-based company was the largest in Britain during the 1820s and supplied textile machinery all over the British Isles and even sent a waterwheel to America. Two of their waterwheels, dating from 1823–4, survive at Thwaite Mills, Leeds, and, although these are not of the suspension type, they are predominantly of iron with some unusual constructional details.

Although iron waterwheels had several advantages in that they could be more rapidly assembled and lasted longer than their timber counterparts, there were several drawbacks with suspension wheels, such as getting the structure so that the ring gears ran truly, in order to prevent excessive noise and wear of their iron teeth. Hewes's designs were modified and improved

Two of the four suspension waterwheels designed by Fairbairn in 1830 for James Finlay's Deanston cotton mills, Perthshire.

Remains of a bullock-hide shuttle for a high breastshot wheel at Clifford, West Yorkshire. The wooden roller across the width of the pentrough carried a leather screen that was rolled down to allow water to enter the buckets of the wheel through the cast-iron guide vanes. Probably of mid-nineteenth-century date.

during the 1820s by the Scottish engineer William Fairbairn (1789–1874), who, after an apprenticeship as a millwright, came to Manchester to work for Hewes as a draughtsman in 1817 but left after disagreement later that year. Fairbairn subsequently went into partnership with James Lillie, building waterwheels and textile machinery, initially with the assistance of James Murphy, 'a muscular Irishman', who turned the lathe in their small workshop. In the 1820s Fairbairn was consulted by James Finlay & Company of the Catrine Cotton Works in Ayrshire, Scotland, for whom he built two of an intended four 50 foot (15.2 metre) diameter waterwheels, each capable of producing 120 horsepower (90 kW). The installation cost £4500 and, in his partly autobiographical *Life* published in 1877, Fairbairn claimed that since the waterwheels 'were started in June 1827; they have never lost a day since that time, and they remain, even at the present day, probably the most perfect hydraulic machines of the kind in Europe.' They were dismantled to be replaced by water turbines in 1947. Shortly after the construction of the Catrine wheels, Fairbairn and Lillie were engaged to build a further eight for the same company at the Deanston Works in Kilmadock, Perthshire, powered by the river Teith. Again the original design was not completed, only two wheels being built by Fairbairn and two more, subsequently, by James Smith. The wheels were 36 feet (11 metres) in diameter, each producing 90 horsepower (67.5 kW), and were dismantled in 1949. The Catrine and Deanston wheels were all high breastshot wheels of the suspension type, 'from the increased facilities which a wheel of this description affords for the reception of the water under a varying head', and used a modification of Rennie's sliding hatch, where water was admitted through guide vanes in the end of the iron pentroughs, with adjustable shutters made of hide that could be rolled up or down to allow the maximum head of water to be used.

Among the improvements that Fairbairn made to Hewes's designs for suspension wheels were the substitution of iron wedges for large nuts, to tension and hold the inner ends of the arms securely, and an improved form of bucket. With overshot wheels the problem of air becoming trapped in the buckets as water entered, thus reducing their holding capacity and the power of the wheel, could be remedied simply by reducing the width of the launder or pentrough so that the water spread out as it entered the wheel. With breastshot wheels turning against close-fitting breastwork this was not feasible, so Fairbairn introduced ventilation slots across the width of the wheel between each bucket and sole plate, through which air could escape, thus allowing the buckets to fill more fully with water. As with all Fairbairn's improvements, his design relied on the use of iron. Rolled iron sheet was used for the buckets and sole plates and the first wheel to be constructed with ventilated buckets was at Wilmslow, Cheshire, in 1828. It is claimed that between 1820 and 1851 Fairbairn built some three hundred waterwheels, some of which were exported. Fairbairn's last waterwheel, and the only one known to survive in England, is that built for the flax-spinning mill of John and George Metcalfe at Glasshouses, Nidderdale, North Yorkshire, in 1851 at a cost of £1000. This wheel, which is 24 feet (7.3 metres) in diameter by 21 feet (6.4 metres) wide, was dismantled in the early 1980s and moved across the Pennines to be rebuilt in the wheelpit that formerly contained a Hewes suspension wheel at Quarry Bank Mill, Styal, Cheshire. Much of Fairbairn's work on waterwheels and mill machinery, including the redesign of many components of the traditional corn mill in iron, is summarised in his *Treatise on Mills and Millwork*, first published in 1861–3, which is one of the most significant Victorian engineering texts on water power.

The development of windmills

The most important developments concerned with improving the efficiency and output of windmills in the second half of the eighteenth century were connected with the design of windmill sails and the way in which they were turned to face the wind. Traditional cloth-set *common sails* were gradually replaced in many parts of Britain by *shuttered sails*, in one form or another, and the method of winding mills manually by tailpole was also superseded, but these developments took over fifty years. The first step was Edmund Lee's 1745 patent for a self-regulating wind machine. Lee, an enigmatic character, was a smith working at an iron forge near Wigan, Lancashire, and his patent was a double first in that he proposed a remote control for windmill sails and a self-winding arrangement. He was attempting to answer the problem of the effect of sudden high gusts of wind that all windmills face. His design for sails

Fantail mounted on the steps of a post mill, Aythorpe Roding, Essex.

consisted of pivoting wooden frames connected by chains to a lever on which weights were hung to hold them face to the wind. When the weights were removed or when the wind pressure overcame their effect, the sails would feather and act as air brakes. Lee's patent drawing shows a small tower mill with a small-diameter wind wheel, a *fantail*, which is at right angles to the sails and is geared down to a travelling wheel that moved around the base of the tower. The position of the fantail as shown on Lee's drawing would have been too low and close behind the body of the mill for it to be fully effective and it was not until the late eighteenth century that Lee's idea became a practical reality, perhaps partly due to John Smeaton. The sails of both post and tower mills were originally turned to face the wind by the miller pushing round the body or cap of the mill using a tailpole, an extended timber lever. Some tower-mill caps were turned by winches operated from inside the cap or from ground level by an endless rope or chain. As towers became taller, however, winding became more of a problem, and timber stages were introduced, forming a platform around the tower at a suitable height for setting the sails and operating the winding gear. In Smeaton's design for Chimney Mill, Newcastle upon Tyne, of 1782, a five-bladed fantail is shown, mounted on an almost horizontal frame projecting from the rear of the cap, and similar forms are known from other late-eighteenth-century illustrations. The position of these early fantails was still too low to be fully effective and by the early nineteenth century they were placed on shorter, higher frames. Fantails were also added to post mills in various ways, such as being geared down to a toothed ring or to a pair of wheels that ran on a track laid out around the mill. Iron gearing formed the basis of Lee's patent of 1745, undoubtedly connected to his occupation as a smith, and the development of the fantail and remote-controlled sails relied on small forged and cast-iron components to make them practical.

Smeaton's experiments into windmill sails, made in 1752–3 and carried out using models as there was no reliable method of creating artificial wind, led him to the conclusion that five sails performed better than the usual four. Multi-sail mills were already in existence in the early eighteenth century; one

Smeaton's model for testing the performance of windmill sails. By pulling the cord, Z, the horizontal arm was rotated, causing the sails, G, to turn. Weights put in the basket, P, running on the uprights, TS, enabled the effect of different forms of sail to be measured.

of the two tower mills at Devizes, Wiltshire, is illustrated with six sails and, according to Daniel Defoe, there was a six-sailed wind pump at Islington, in London, for example. Four sails was the common arrangement, however, as two stocks could be mortised through the end of a windshaft and still leave enough timber for secure fixing. Smeaton overcame the practical problem of fitting an odd number of sails by designing a cast-iron cross with five arms that could be attached to the end of an iron windshaft. The earliest known drawing of such an arrangement is contained in his designs for a flint-grinding mill at Leeds in 1774. Smeaton's sails were also wider at the tips than at the heels, giving a triangular shape to the leading edge, a form that does not appear to have been widely adopted, and no examples survive, partly because of the vulnerability of windmill sails and hence their frequent repair and replacement. All of Smeaton's designs show cloth-set sails and he also worked out the *weather* or twist of the sail frames to give the best performance, although his experiments confirmed that the form that had already been developed empirically by millwrights was about the best.

The development of shuttered sails was begun by

Great Gransden post mill, Cambridgeshire, in the 1890s, showing a pair of common sails set with canvas and a pair of spring sails with the shutters open.

Chillenden, Kent, showing four spring sails with the shutters open. The springs controlling the shutters on each sail can be clearly seen.

Andrew Meikle (1719–1811), a practical and inventive millwright who lived and worked on the Phantassie estate of the Rennie family at East Linton in Scotland. Meikle proposed sails with the frames divided into a series of bays containing shutters of wood and canvas pivoting on iron rods. The movement of the shutters was controlled by pre-tensioned springs, which allowed them to open and spill wind when its force overcame the pressure of the springs. Smeaton suggested improvements such as connecting all the shutters on each sail together with longitudinal bars rather than trying to control each shutter separately, and the first *spring sails* were built in about 1772. They do not appear to have become widespread until the middle of the nineteenth century, however, and tended to be used mainly on smaller tower and post mills, partly because of the need to stop the mill to set each sail individually. Often a pair of spring sails was put up with a pair of commons, which gave a good combination of driving power and self-regulation.

Some five patents were taken out for different types of remote-controlled windmill sails between 1785 and 1807, most of which were short-lived in practical terms. An exception was Captain Stephen Hooper's patent of 1789, which proposed the use of shutters in the form of canvas roller blinds that were connected together and controlled remotely from within the mill. After some practical modifications the design became known as the

Hooper's roller reefing sails on the tower mill at Tollerton, North Yorkshire, in 1928.

roller reefing sail and the V-shaped air poles that linked the shutter bars to the striking gear, by which the rollers were furled or unfurled, are quite distinctive, although no examples now survive in England. The tower mill at Ballycopeland, County Down, retains them in a modified form, however. Hooper's sails were put up on two windmills at Deal in his home county of Kent before 1791, but the design spread most rapidly in East Yorkshire, through the activities of the Hull millwrights Norman & Smithson. A five-sailed wind oil mill built for William Dibb at Stoneferry, Hull, in 1791 was one of the first to be fitted by them with Hooper's sails, and they continued to use his sails, under licence, until the introduction of a superior self-regulating form after 1807.

The *patent sail*, as it has become known, was the design of the engineer William Cubitt (1785–1861), of Bacton Wood Mill, North Walsham, Norfolk, who combined the rigid shutters proposed by Meikle, be they of wood and canvas, board or metal, with a control or striking gear modified from Hooper's design. Cubitt worked out an appropriate remote-control mechanism where the shutters on each sail were connected together by a rod, which in turn was linked by cranks, the *spider*, to a *striking rod* passing through the length of the windshaft. The windshaft had to be bored through or cast hollow to accept the rod, the position of which could be altered by either gearing or a lever, so that the sail shutters could be opened or closed by pulling an endless chain that hung down behind the mill body. Weights hung on the chain kept the shutters closed until the force of the wind overcame their effect, and the sails could be easily regulated without stopping the mill. The first windmill to be fitted with Cubitt's sails was a smock mill belonging to his father-in-law, Samuel Cooke, at Stalham, Norfolk. The sail design was patented in 1807 and could be used only by millwrights under licence until 1821. Subsequently the patent sail became the most widespread form used in England.

Although windmills continued to be built throughout England during the nineteenth century, the zenith was achieved before 1830. It can be seen in mills such as Alford and the Maud Foster Mill, Boston, Lincolnshire, which was designed and built by the Hull millwrights Norman & Smithson for

The zenith of English windmilling: Alford, Lincolnshire. Dating from the first quarter of the nineteenth century, the brick tower mill has five single-sided patent sails and a distinctive ogee cap, winded by fantail.

Isaac and Thomas Reckitt in 1819 at a cost of £1200, excluding the brickwork of the tower. The tall brick tower contains seven floors and is crowned with an ogee-shaped cap winded by fantail. The mill has five patent sails mounted on a cast-iron cross, and iron shafting and gearing take the drive to three pairs of millstones. The original design drawings show a cylindrical grain cleaner on the floor above the millstones and a flour dresser suspended from the ceiling of the ground floor, to make best use of gravity for feeding grain and meal through the milling process, the finished product ending up conveniently at ground-floor level. It was the millwrighting firms such as Norman & Smithson which carried forward the inventions and designs of the eighteenth-century English and Scottish engineers and built the fine mills that mark the peak of development. But at the same time as Norman & Smithson were completing what was probably their last windmill in Boston, many millwrights were still building water- and wind-powered machinery using traditional materials and methods, for both practical and economic reasons. At Bursledon, Hampshire, for example, a brick tower mill with four common sails, hand-winded cap and all-timber gearing and shafting driving two pairs of stones was completed in 1813–14, its design and construction reflecting the sturdy, traditional millwrighting that had been current in the mid eighteenth century.

Land drainage

The eighteenth century was the great age of windmills for fen drainage but, compared with the developing forms of windmill found elsewhere in England, drainage mills were relatively modest, for reasons of economy and convenience. They were generally smock mills, sometimes raised on brick bases, built and geared in timber with four common sails and tailpole winding. In the Middle Level of the fens around Manea, Cambridgeshire, there were some two hundred and fifty drainage windmills in the mid eighteenth century, but their building and use were not without difficulty, as local nuisance was often caused by water being drained from one area only to cause problems elsewhere. John Rennie, who was much involved with drainage schemes in the first two decades of the nineteenth century, described wind-

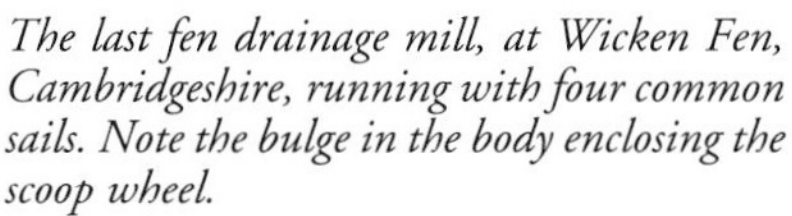

The last fen drainage mill, at Wicken Fen, Cambridgeshire, running with four common sails. Note the bulge in the body enclosing the scoop wheel.

mills as a 'very imperfect and expensive mode of drainage; especially when wet weather is succeeded by calm weather, the mills cannot work, and therefore the water lies on the surface of the Fen, and does incalculable injury.' His proposals to use steam power were largely ignored, the drainage boards preferring to spend small sums on the repair and maintenance of existing windmills rather than investing in steam, although one large steam engine could do the work of four or five windmills. The first steam engine for land drainage was probably that erected at Hatfield Chase, Lincolnshire, in 1813, and by 1820 steam engines were at work in the Fens.

Steam power

After the successful introduction of Thomas Newcomen's atmospheric steam engine in the second decade of the eighteenth century, the improvement and efficiency of this new prime mover became the concern of millwrights and engineers, including James Brindley, John Smeaton and, in particular, James Watt. The reciprocating action of the early engines was suitable for pumping water but rotary motion still depended on waterwheels until the 1770s, so a compromise was to build a waterwheel-driven plant and to supply the water to the wheel by a steam pump, known as a *returning engine*. There are a number of early examples, significantly at Matthew Boulton's Soho factory in Birmingham, where water from the Hockley Brook was pumped on to a 20 foot (6 metre) diameter overshot wheel used for rolling metal for plated ware and also for providing power for the mint. A returning engine was used at the Carron Ironworks in 1780, and Smeaton also proposed such an arrangement to His Majesty's Victualling Office at Deptford in 1781, as he doubted that steam power would be as even or as steady as that of water for driving millstones. The first steam-powered corn mill using direct engine drive, rather than pumped water, was built by Young & Company in Lewins Mead, close to the centre of Bristol, in about 1779, using a steam engine with a system of ratchets, built to the design of Matthew Wasbrough. In 1780 James Pickard, who collaborated with Wasbrough, patented the application of the crank to the steam engine, and it was the restriction of this patent that encouraged James Watt to invent his sun and planet gearing, which proved a significant success and allowed steam power to be widely used where continuous rotary motion was required. Boulton & Watt made the steam engines for the Albion Mills, Blackfriars, London, which were built to the designs of the architect James Wyatt, with iron millwork by John Rennie. Although they were never fully completed and burnt down within five years of starting work in 1786, the Albion Mills represent the first attempt to turn grain milling into a factory process, an idea that was to take nearly a century to develop.

Corn milling

Grinding grain to produce meal and flour was always the major use of windmills in Britain and, in the late eighteenth century, in spite of improved waterwheels and the setting up of steam mills, the milling industry came to

Sarre Mill, Kent, a fine smock mill built by the millwright John Holman of Canterbury in 1820 and now restored to full working order.

rely more and more upon windmills. There was a significant increase in the number built during the first quarter of the nineteenth century, owing to the demands of an expanding population and also to the pressure put on people and resources by the Napoleonic Wars. In Kent, for example, the number of windmills rose from about a hundred in the mid eighteenth century to nearly two hundred and fifty by the middle of the nineteenth. In some urban and industrial areas windmills were built as investments, where, although there was adequate water power, it was being used for other functions. In about 1800 Thomas Smart, a baker, built a tower mill overlooking the town of Bradford-on-Avon in Wiltshire, the river Avon below it being used intensively at that time for powering woollen mills. Tower mills were also built in the cotton town of Belper, Derbyshire, in 1796 and at Croxley, Hertfordshire, after 1820, where the nearby rivers Chess and Gade were busy driving paper mills. As in the twelfth century, when it had first been harnessed, wind power was used to supplement that of water.

A small number of mills were built or extended during the early nineteenth century to combine both wind and water power. In most cases, windmills were added to existing watermills, and the evocative name of Doolittle Mill, at Totternhoe, Bedfordshire, suggests

Combined windmill and watermill at Legbourne, Lincolnshire. The windmill, built in 1847 by Saunderson of Louth on an existing watermill site, worked into the 1920s. It is now a house.

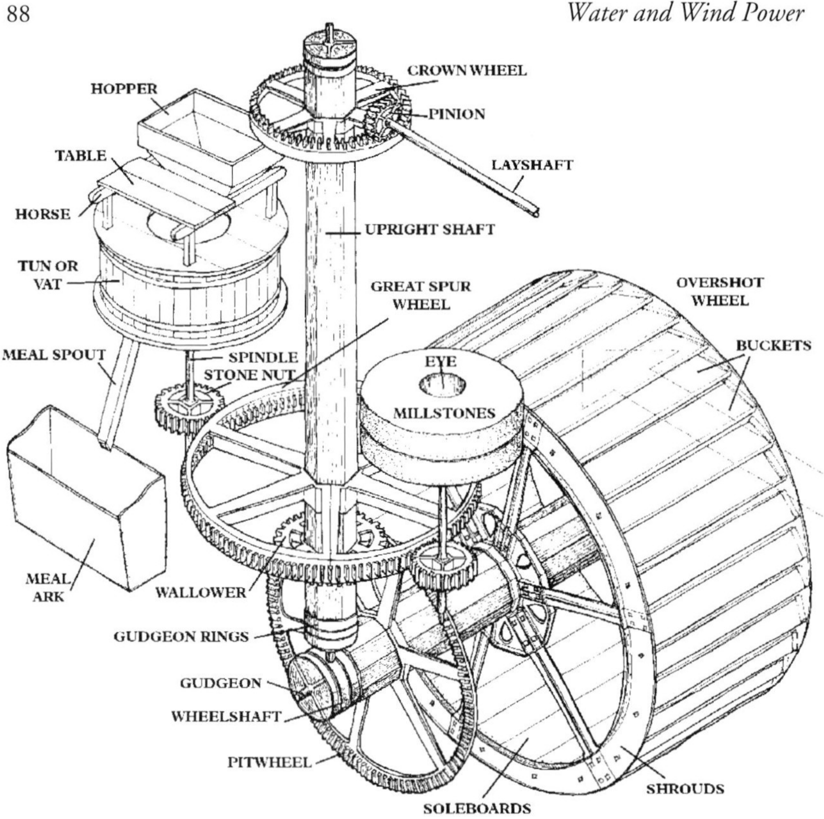

Typical spurwheel drive layout in a corn mill, showing an overshot waterwheel driving two pairs of millstones, based on Felin Newydd, Powys.

a good reason for the need to combine the elements. At Bishopstone and West Ashling, Sussex, windmills were built on the roofs of tide mills to power the sack hoists so that work could continue regardless of the state of the tide, and there were a number of windmills separate from but closely associated with tide mills, for example at Thorrington and Walton on the Naze, both in Essex.

The development of spur gearing and the general improvement of both waterwheels and windmill sails allowed larger mills to be built, driving more pairs of millstones, as well as ancillary machinery for hoisting, cleaning grain and dressing flour. John Smeaton's designs for watermills at Halton, Lancashire, and Wakefield, West Yorkshire, made in 1754, both show three pairs of millstones and spurwheel drive. By the 1770s this was a fairly common arrangement, with the upright shaft also carrying a crown wheel from which layshafts were driven to work a sack hoist and dressing machinery. The use of more than one type of millstone also became common, and a survey of Keighley Mill, West Yorkshire, made in 1772 records grey, blue and French stones and

Millstone governor mounted on the spindle below the stone nut at Norton Lindsey, Warwickshire. The steelyard, which alters the gap between the stones, connects the governor to the bridge tree, top right.

a gear-driven flour dresser. Grey stones are monolithic stones of millstone grit, mostly quarried on the Pennines near Sheffield on the Yorkshire–Derbyshire border. They were used for general grinding and, latterly, animal feed. Blue stones, also known as 'blacks' or 'cullins', are of a German lava that was imported in the form of querns and millstones from Roman times, and their use persisted in the north and east up to the end of the eighteenth century. They were mainly used for flour and malt milling. French stones or 'burrs' were imported into England from the sixteenth century, usually in blocks from which millstones were built up, backed with plaster of Paris and bound with iron hoops. In a list of merchandise brought out of France into England in about 1604 millstone and plaster of Paris are listed consecutively, implying a connection at that early date. Because of their hardness and ability to cut the bran cleanly from wheat grains during milling, French stones became the most widespread in British mills.

The problem of maintaining automatically the critical balance between the speed of the mill and the distance between the millstones, so that millers could control the texture of meal and thus the fineness of flour, was also solved during the second half of the eighteenth century with the introduction of governors or regulators. The problem was particularly acute in windmills, as the speed of a waterwheel can be more readily controlled by adjusting the water feed. Several patents were taken out, notably by Thomas Mead in 1787 and Stephen Hooper, as part of his design for the roller reefing sail in 1789. It

Gear-driven wire machine at Felin Newydd, Crugybar, Carmarthenshire. The hopper, top right, feeds the ground meal into the ribbed cylindrical sieve, through which the fine flour is brushed.

seems that neither was completely original, but the use of governors to maintain the distance between the stones regardless of variations in wind speed was quickly taken up by millwrights and millers, and most surviving windmills have them. They were used by Rennie in the Albion Mills by 1786, and in 1788 James Watt adapted the device to control the throttle valve of his steam engines. Governors were also used to control the shutters of windmill sails, but the widespread use of Cubitt's patent sails from the second quarter of the nineteenth century tended to render them obsolete for this application.

There were also important developments in the ancillary machinery used for cleaning grain and mechanically sifting the meal after milling to extract flour. The introduction of winnowers and *threshing machines* resulted in better cleaning of grain after harvesting, and John Milne's patents of 1765 and 1771 for flour-dressing machines using sieves of woven wire mesh improved both the quantity and the variety of products that could be made, although the traditional bolter with its more gentle action still produced a better colour flour. The introduction of more machinery into mills resulted in the development of mill buildings. A three- or four-floor layout allowed grain to be stored, cleaned, milled and dressed within the same building, making use of a power-driven hoist to lift the sacks to the upper levels and then gravity to feed it through spoutwork to the machines and millstones. In Scotland and parts of upland England, specialised machinery for preparing pearl or pot barley and for shelling and grinding oats was also developed, with an increasing use of rope and belt drives for the lighter, faster-moving ancillary equipment.

Spurwheel drive became the most widely used gearing system in British mills, both wind- and water-powered, although there are many interesting variations. In windmills the millstones are generally driven from above, *overdrift*, and in watermills from below, *underdrift*, but there are examples of mills where the systems are reversed for a variety of reasons, such as the influence of a particular millwright or an adaptation to suit local conditions when the millwork was rebuilt or modernised. It is perhaps sufficient to say that if a gearing layout could exist it probably does somewhere, and it is the variety and individuality of design that make corn mills such a broad and fascinating study. The idea of horizontal windmills was also pursued during the eighteenth century, but they continued to be a curiosity rather than a practical alternative. Archaeological excavation and historical research have identified the sites of three horizontal windmills for corn milling built by Thomas Mortimer near Eastbourne, Sussex, in the mid eighteenth century. Captain Stephen Hooper, patentee of the roller reefing sail and governor, built horizontal windmills at Margate and Sheerness, Kent, and also designed the large mill built by Thomas Fowler at Battersea in London in 1788 (originally for producing oil and subsequently used for corn milling), which was dismantled in about 1825.

The management of corn mills also underwent considerable change during the last decade of the eighteenth century, owing to social and economic pressures. In 1796 Parliament passed the Act for the Better Regulation of

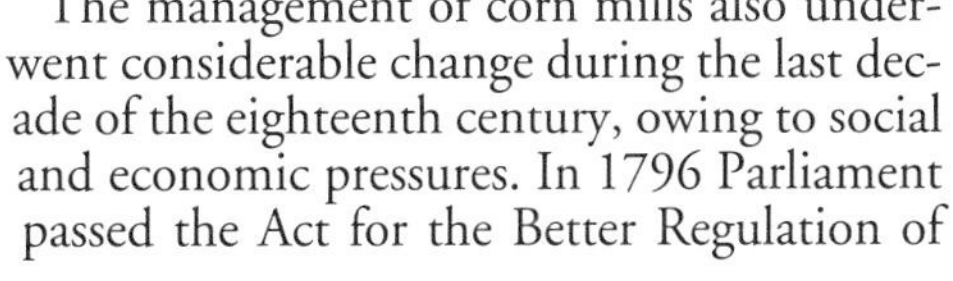

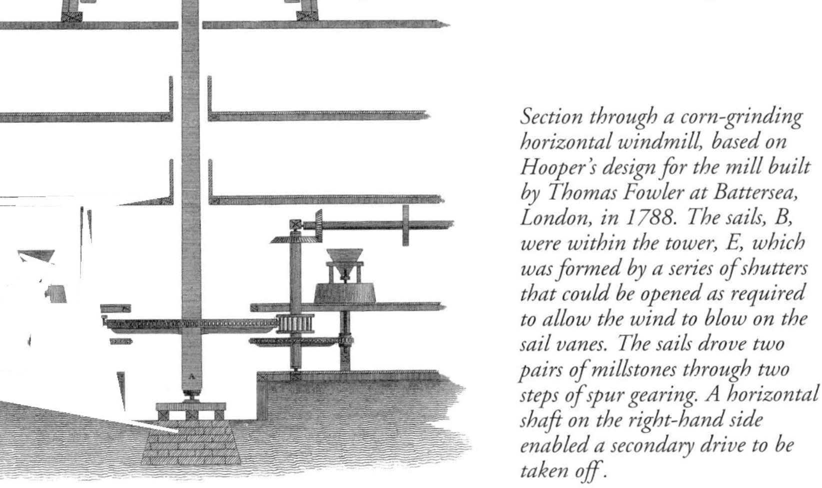

Section through a corn-grinding horizontal windmill, based on Hooper's design for the mill built by Thomas Fowler at Battersea, London, in 1788. The sails, B, were within the tower, E, which was formed by a series of shutters that could be opened as required to allow the wind to blow on the sail vanes. The sails drove two pairs of millstones through two steps of spur gearing. A horizontal shaft on the right-hand side enabled a secondary drive to be taken off.

Mills, which regulated the miller's toll and encouraged the replacement of the toll-in-kind system with a monetary one. During the 1790s there was considerable fluctuation in the price of grain and bread because of bad harvests and the wars with France, and a number of union or subscription mills were built throughout Great Britain, particularly in industrial areas. These mills were an early form of co-operative venture, being financed by subscriptions raised by shareholders for the purpose of purchasing grain and making it into flour and bread for distribution to the subscribers at prime cost and also to prevent the adulteration of flour. Wind power was well suited to serve these relatively small-scale ventures, although water and steam were also used. One of the earliest enterprises was the Anti Mill Society, founded by the 'poor inhabitants' of Hull in 1795, for whom a tall five-sailed tower mill was built in 1796, probably by millwrights Norman & Smithson, as was a similar mill at Whitby in 1801. Many others were built during the first two decades of the nineteenth century, and the tower mill at North Leverton, Nottinghamshire, founded as a subscription mill in 1812, is still run by a company formed by descendants of the original shareholders.

The siting of windmills was restricted by the Turnpike Roads Act of 1822, which stated that 'No person shall hereafter erect or cause any windmill to be erected within the distance of 200 yards [183 metres] from any part of any turnpike road.' There are several instances where windmills were built on new sites to replace existing mills, as at Shapwick, Somerset, where a post mill close to the junction of two turnpike roads was blown down in 1836. It was replaced by a stone tower mill on a new site some distance away, the 'great nuisance' of the old mill having been removed.

The number of pairs of millstones in a windmill was generally limited by the space available, with three or four pairs being the most practical in a large tower or smock mill. In watermills the limitation was based more on the availability and consistency of water power but both windmills and watermills could lie idle for prolonged periods because of the vagaries of the weather. A small corn mill in a rural area may have been built with two or three pairs of stones, with hoisting, grain-cleaning and flour-dressing machinery, but probably only worked to capacity during the autumn and winter months when water levels were highest. In 1815 the Navigation Mill at Warwick, which was turned by the superfluous water from the Warwick and Napton Canal, drove five pairs of stones, with 'three constantly at work . . . and capable of grinding and dressing for bread flour, upwards of 300 bushels [7.5 tonnes] per day'. A number of large millstone plants were constructed, notably in the 1790s at Warrington, Cheshire, where the mills designed by the Yorkshire millwright John Sutcliffe had nineteen pairs of stones.

Industrial wind and water power

The many uses to which windmills were put in the early industrial period are vividly illustrated in the Dutch mill books published during the first half of the eighteenth century. The situation in Britain was different, however, for waterwheels were the most reliable, effective and economic prime movers throughout the eighteenth century, and windmills used for purposes other than milling corn and pumping water for land drainage have always been a small but interesting minority. Probably the greatest concentration was in the Hull area towards the end of the eighteenth century, where wind power was used to drive mills for oil extraction, sawing timber, making paper and grinding whiting for putty and paint, an industry that also required oil. The influence of Yorkshire millwrighting spread through the activities of industrialists and entrepreneurs such as Samuel Walker, a Yorkshire ironmaster, who built two tower mills for grinding white lead at Islington, London, in 1786 and 1792. These mills probably represent the earliest use of five sails and fantails in the south of England.

A small number of wind- and water-powered machines were developed during the eighteenth century that reflect contemporary developments and thought but were essentially short-lived and are now seen rather as curiosities. In 1752 a little eight-sailed windmill, in appearance rather like Lee's fantail, was designed by Thomas Yeoman and built by the millwright Cowper or Cooper of Poplar, London, to drive ventilators at Newgate Prison in London. The ventilators were the idea of the Reverend Dr Stephen Hales, whose work on plant physiology led to the belief that fresh air was essential to prisoners and also common seamen. Windmills were also set to work on warships and at other jails. The Coopers were a millwrighting family who subsequently undertook work for John Smeaton and also had a virtual monopoly of brewery millwrighting in London. In 1787 James Cooper patented a newly invented watermill that comprised a series of flat floats fastened to an endless chain that ran over a wheel, from which the drive was taken. Water admitted through a sluice gate passed through an aperture through which one side of the chain of floats also ran, so that the fall of water turned the device by gravity. Cooper claimed his invention was particularly useful for tide mills, where the head of water during working was continually falling.

Wind power was also adopted in a number of industries where water power was already being utilised to capacity. At William Champion's brass works at Warmley, on the outskirts of Bristol, a tall stone tower mill built in about 1750 seems to have been originally intended to work pumps for recycling water to power waterwheels in the battery mill, but in 1761 it was described as 'One windmill with Stamps &c' and was used to crush calamine (zinc ore) for the preparation of brass. As with other industrial windmills, the processing machinery appears to have been located in buildings surrounding the base of the tower. Smeaton's designs of 1755 for a wind oil mill at Wakefield, West Yorkshire, show an arrangement reminiscent of some illustrated in the Dutch mill books, where a timber-framed tower is simply a supporting structure for the cap and sails, and the edge runner stones and stamps used for extracting oil are housed at ground level in a square brick base. Smeaton carried this design over into corn milling with his later designs for Chimney Mill, Newcastle upon Tyne, where the potential of using the tower for storage and gravity feed of grain for milling is not realised.

Water power had been used for grinding oak bark for tanning leather from medieval times and by the sixteenth century was also used for preparing dyestuffs for the textile industry, which, until the introduction of chemical dyes in the nineteenth century, were extracted from plants and wood. A mill on the river Wandle at Wandsworth, London, was producing dyestuffs before 1580 and by the mid eighteenth century had become a combined windmill and watermill, a smock mill being raised on top of the watermill. A similar mill, referred to as a 'Brazil mill', existed at Hounslow, Middlesex, and a drawing of 1757 shows a six-sailed smock mill raised on a three-storey square base, similar to other contemporary industrial windmills. Brazil-wood and logwood were two raw materials commonly used in dyeing, being imported in the form of small logs from Central and South America. The logs were reduced to a coarse powder by rasping in special machines, or by chipping the wood using a circular disc with inserted blades and then crushing the chips under edge runner stones.

Wood-turning mills were once common in the Lake District, where local coppice wood was used to make *bobbins* for the textile mills of Lancashire, Yorkshire, Scotland, Ireland and also abroad. A small watermill for turning bobbins was advertised at Staveley, Cumbria, the centre of the industry, in 1797 and by the middle of the nineteenth century there were about forty-nine mills at work. At Stott Park, on the west side of Lake Windermere, a bobbin mill built in 1835 was originally driven by a waterwheel of about 32 feet (9.7 metres) in diameter, later superseded by water turbines and a steam engine.

Sawmills

Water power was used in northern Europe from late medieval times to drive mills for sawing deal and planks using reciprocating saws mounted in vertical frames driven by cranks, the action of which imitated pit sawing. The first wind-powered sawmill was built in the Netherlands in 1592 and it was a Dutchman who built the first wind sawmill near London in 1663, which was abandoned after a riot by sawyers who feared that it would destroy

Water-powered sawmill with reciprocating frame saw, from John Evelyn's 'Sylva', 1670.

their livelihood. In 1767 a wind sawmill was built at Limehouse, London, with encouragement from the Society of Arts. A Mr Dingley supplied the capital, the mill costing £4454 to build, and James Stansfield, a carpenter from Bingley, West Yorkshire, provided the technical knowledge. The mill drove thirty-six saws, rather more than the Dutch mills on which its design was based. Like its predecessor a century before, it was attacked by a mob and put out of action for six months, costing £1231 to repair. Two sawmills, one water- and one wind-powered, were built at Garmouth, Strathspey, in about 1786 by a company owned by Mr Dodsworth of York and Alderman Osbourne, a timber merchant, of Hull, who had purchased the forest of Glenmore and transported timber from there to Hull and the naval dockyards at Deptford and Woolwich. The windmill was reported to have contained between thirty-six and forty saws. Osbourne also owned a wind-powered sawmill in Hull in 1798, perhaps built by Norman & Smithson, which was the 'first ever seen in the area' and in 1819 was described as having eight patent sails.

A circular saw invented by Samuel Miller of Southampton in 1777 was used by William Taylor in 1781 in his workshop at Woodmill, on the river Itchen, where he set up an elaborate plant driven by a waterwheel to supply the Admiralty with ships' blocks. The accurate cutting and boring of the timber cheeks that enclosed the sheaves over which ropes ran meant better action and thus the ability to alter sail more quickly, without blocks jamming, which was one of the advantages that Nelson's fleet had over the French at Trafalgar. Taylor's workshop was thus a vital early step in the development of machine-tool production. There were few sawmills in England until the 1820s, however, although there were over sixty water-powered mills in Scotland by 1830, converting both home-grown and imported timber. At Gunton Park, Norfolk, a timber-framed and thatched sawmill, which dates from the 1820s, has two overshot waterwheels fed from a pond filled by a tributary of the river Bure. One wheel latterly drove a circular saw, the other a vertical reciprocating frame saw, which has been restored to working order. Many Victorian estates had water- or engine-driven sawmills for converting timber for their own use; some of these were purpose-built and others were put into existing buildings. At Fountains Abbey the medieval corn mill was converted to an estate sawmill and a new waterwheel was installed in a brick wheelhouse in the early nineteenth century, and at Dunham Massey, Cheshire, an early seventeenth-century corn mill was turned into the estate sawmill and turning mill in about 1860.

Metalworking

As well as being used in mining and industrial applications, there were many ancillary processes and trades that relied on water power. In 1770 there were 161 separate works driven by waterwheels on the five rivers of Sheffield and their tributaries, many of which were engaged in making forgings, tools and cutlery. Edge-tool works used waterwheels to drive bellows for hearths, tilt or trip hammers for forging and *grindstones* for finishing and sharpening. At Abbeydale, on the river Sheaf to the south of Sheffield, the dam (synonymous

Water-powered forge with two trip hammers and shears at Finch Foundry, Sticklepath, Devon, in the 1950s. The camwheel for tripping the lighter, faster hammer can be seen behind the hammer frame.

with pond in Yorkshire) was enlarged in 1777 and the stone building containing the tilt forge is dated 1785. It has a high breastshot wheel driving the tilt hammers and a smaller overshot wheel working a blowing engine. The grinding hull was rebuilt in 1817; this houses a number of grindstones belt-driven from layshafts driven by a second high breastshot waterwheel.

Needle making started at Studley and Alcester in Warwickshire in the seventeenth century and by 1730 water power from the river Arrow was being used at Redditch, Worcestershire, in a former forge mill, for grinding the points and for *scouring,* the finishing process of polishing needles. Edge-tool production using water power was also concentrated in the west Midlands in the area around Belbroughton, Worcestershire, where scythes had been made since the fifteenth century, and in north Somerset, where James Fussell leased an ironworks in the Wadbury valley, near Frome, in 1744. In 1791 the family operated several ironworks in the area, which were said to supply 'every iron implement of husbandry', including scythes, sickles, spades and shovels. Similar small ironworks and forges existed in the West Country to serve the local agricultural and mining communities, such as that established by William Finch at the former Manor Mills at Sticklepath, on the north side of Dartmoor, Devon, in 1814. A woollen mill was converted to a forge, with two tilt hammers and shears driven by an overshot waterwheel, and an undershot wheel was installed in the tailrace to drive a fan for providing a draught for the hearths and furnaces in about 1835. The nearby manorial corn mill was subsequently taken over and converted into a grinding house for finishing the tools.

Stone

'Black marble', a form of limestone, was used locally in Derbyshire for church furnishings from medieval times, and the first water-powered marble mill in England was built by Henry Watson at Ashford, near Bakewell, in 1748. The stone was cut by vertical banks of iron sawblades and then taken to a sweeping mill, where the slabs were ground to give an even surface and thickness. Both processes were powered by waterwheels driving machinery by cranks. Wind and water power were also used to drive machinery for grinding stone, mainly for making glazes for the pottery industry. There were a number of water-powered flint mills in north Staffordshire, built to serve the pottery industry which developed there from the seventeenth century. Flint was originally ground using conventional millstones and stamps, but these methods created dangerous amounts of dust. In the early eighteenth century wet grinding was developed, using water power, at Hanley, near Stoke-on-Trent. The mills worked on the principle of a circular grinding pan into which batches of calcined flint were put with water, to be pushed around with loose blocks of chert (a flinty mineral) known as *runners* by sweep arms connected to a vertical shaft. The base of the pan was also made of chert and, as the arms rotated, the flint was worn down to form a slurry with the water. When the grinding was finished, the slurry was run off and the water evaporated, usually in kilns, leaving the ground flint to be made into blocks that were then used by potters for making glazes. Bone, for mixing with clay to make fine china, was ground in the same way. There was a concentration of flint-grinding mills in the Moddershall valley, Stone, Staffordshire, some having been converted from other uses in the mid eighteenth century. At Cheddleton, Staffordshire, a corn mill on the river Churnet was converted to flint grinding and a second mill was built on the north side of the stream sometime before 1783, both driven by low breastshot waterwheels. Flint-grinding mills were also built to serve the potteries of Newcastle upon Tyne and Leeds, and John Smeaton's design for a five-sailed flint-grinding windmill in 1774 has already been mentioned. The only windmill known to have been built by James Brindley was for John and Thomas Wedgwood at Burslem, Staffordshire, in about 1750; it drove a single pan for grinding flints.

The use of Cornish stone and china clay to form a porcelain body was discovered by William Cookworthy of Kingsbridge, Devon, in the mid eighteenth century, and a number of water-powered stone-grinding mills were built in the china-clay area around St Austell, Cornwall, during the nineteenth century. The industry benefited from the expertise already established in mining and quarrying, and surplus machinery was utilised in the china-clay workings as the mining industry waned later in the nineteenth century. Many pumping and winding waterwheels, and the knowledge of those who built and worked them, thus survived into the twentieth century. Many of the tools that were produced in the West Country forges were used in the china-clay industry, as well as in agriculture.

Snuff

Before the beginning of the eighteenth century most snuff sold in Britain was imported but, as the demand grew, both watermills and windmills were adapted to produce this fashionable commodity. The two windmills on the castle mound in Devizes, Wiltshire, were used for grinding snuff in the mid eighteenth century, and there were watermills in west Wiltshire in similar employment. Much tobacco was imported through Bristol, where tower mills at Clifton Down and Cotham were converted to snuff mills in the second half of the eighteenth century. In Sheffield the Wilson family started grinding snuff using water power at Sharrow Mills on the Porter in the 1740s, the prepared tobacco leaf being pulverised by large banks of iron pestles rotating in oak-lined mortars. Snuff-taking remained a popular habit until Victorian times, when smoking was considered to be more elegant.

The lower dam and the top of the tower that contained three overshot waterwheels at Lumbutts Mill, Todmorden, West Yorkshire.

THE NINETEENTH AND TWENTIETH CENTURIES

In the first edition of his *Treatise on Mills and Millwork* (1861) William Fairbairn devoted only four pages to windmills and summarised his reasons thus: 'Now both sources of power [water and wind] are also abandoned in this country, having been replaced by the all-pervading power of steam. This being the case, we can only give short notice of wind as a motive power, considered as a thing of the past.' Fairbairn also commented on the change in the character of rivers, due to an extended system of drainage that was favourable to land and agriculture but not to large-scale water-power use. The continuing growth of urban areas and the supply of water for domestic needs, as well as the requirements of an expanding network of canals, were rarely compatible with the needs of water-power users and had a noticeable effect. However, even in the second quarter of the nineteenth century, when most of the available water power in Great Britain had already been harnessed and it was impossible to guarantee a water supply in many situations owing to seasonal variations such as drought and frost, it is apparent that some mill owners were still not convinced of the efficacy of steam.

In the Lumbutts valley, near Todmorden, West Yorkshire, the Fielden brothers built a series of new dams in about 1830 to supply water to the cotton mills they had been developing since the 1790s. At the top mill site, formerly occupied by a corn mill, they constructed a stone tower 98 feet (29.8 metres) high, containing three overshot waterwheels one above the other, each 30 feet

(9.1 metres) in diameter by 6 feet (1.8 metres) wide. Water from the Pearson and Healey dams was put on to the top wheel, from the Lee dam on to the middle and from the Old dam on to the lower. The combined water power was estimated to be about 54 horsepower (40.5 kW), but the normal working output was only about half that. The system remained in use into the 1890s, however. In 1835 the Factory Inspectors' Returns for Yorkshire cotton mills indicate that water and steam were both generating about the same amount of power, but the statistics show that, whereas only eighteen engines out of a total of seventy-seven were producing less than 10 horsepower (7.5 kW), there were seventy-five waterwheels out of 134 in the same power bracket. By the mid nineteenth century steam power dominated in the textile industry.

Water power was still widely used in areas where coal was not readily available, such as in the metal-mining districts of south-west and northern England, the Isle of Man and Wales, where large-diameter wheels continued to be used for pumping and winding. Much information about the sizes and construction of these 'water engines' comes from sales particulars, for it was not uncommon for wheels to be dismantled and moved from one site to another as fortunes changed. Pumping wheels could be located at a distance of a quarter of a mile (0.4 km) or more from the site where their work was required, and survivals are rare, although the courses of leats and fine-stone wheelpits can sometimes be found. The largest waterwheel recorded in Cornwall, at Boswedden Mine, St Just, in 1837, was 65 feet (19.8 metres) in diameter. Many others were over 50 feet (15.2 metres) and the greatest concentration was at the Wheal Friendship and Wheal Betsy lead and copper mines near Mary Tavy, Devon, where by 1842 a total fall of 526 feet (160 metres) powered seventeen overshot wheels, eight for pumping, four for winding ore to the surface, and the remainder for crushing and stamping. The largest was the Old Sump wheel, a pumping wheel of 51 feet (15.5 metres) in diameter and 10 feet (3 metres) breast, which was capable of producing 87 horsepower (65 kW). Devon Great Consols were a group of mines in the Tamar valley on the borders of Devon and Cornwall that opened in the 1840s and became the largest copper producer in Europe over the following twenty years. At one time they had thirty-two waterwheels up to 50 feet (15.2 metres) in diameter. At the Coniston Copper Mines, Cumbria, there were at least thirteen waterwheels up to 50 feet (15.2 metres) in diameter at work in 1850. The largest surviving waterwheel in the British Isles, the Lady Isabella Wheel on the Isle of Man, was built for pumping water from the zinc and lead mines at Laxey in 1854. It was designed by Robert Casement and is a pitchback wheel of about 72 feet 6 inches (22 metres) in diameter by 6 feet (1.8 metres) breast, estimated to produce about 200 horsepower (150 kW).

The Victorian era is one of great contrasts in the use of water and wind power. The major industries such as textiles, mining and heavy engineering continued to replace water power with steam and, as will be seen, the incentive for the development of water power for industrial applications passed to France and the United States. But conventional waterwheels and windmills

The largest surviving waterwheel in the British Isles, the Lady Isabella Wheel at Laxey, Isle of Man, built in 1854 for mine drainage.

The high quality of Victorian millwrighting and engineering can be seen at many mills, a good example being this iron breastshot waterwheel made by the Dorchester millwrights Winter & Hossey at Maiden Newton, Dorset.

continued to be built throughout the nineteenth century and there are many examples of their continuing importance, particularly for light industry and grain milling. In 1847 James Coate, the inventor of a cement for holding the bristles in toothbrushes, rather than using wax, took over a former corn mill at Nimmer Mills, Chard, Somerset, and installed machinery for making bone-backed toothbrushes and shaving brushes, having been 'attracted to the Westcountry by the cheapness of water power'. In the 1860s thirty-eight mills on the river Wandle in Surrey were still using water power, assessed at a total of 781 horsepower (586 kW), giving an aggregate of about 20 horsepower (15 kW) at each site. In 1864, when most of the wheels in the Loxley valley, Sheffield, were destroyed in a devastating flood, the majority of works were subsequently rebuilt and continued to use water power into the twentieth century, even though coal was readily available locally. By the end of the nineteenth century iron founders and millwrights were offering waterwheels as catalogue items, for a variety of different uses, including water pumping and agriculture.

Power on the farm

The introduction of machinery for winnowing and threshing grain, and cutting straw, hay and root crops for fodder, during the second half of the eighteenth century encouraged the use of power on farms. At first animal power dominated and the distinctive roundhouses in which horses rotated overhead gears were once a common feature attached to barns, particularly in upland areas of Britain. Waterwheels and, to a lesser extent, windmills were also used. Andrew Meikle, who patented the first successful threshing machine in 1788, described his machine as being 'capable of being worked either by Cattle Wind Water or any other Power' and illustrations of horse-, water- and wind-powered machines are given in Andrew Gray's *The Experienced Millwright* (1804). Water-driven threshing mills became widespread in upland areas, and wind power was also used in the north of England and Scotland. A tall stone tower mill, probably dating from the early nineteenth century,

An early-nineteenth-century windmill tower and later horse wheelhouse beside a threshing barn at Shortrigg, Hoddam, Dumfries and Galloway.

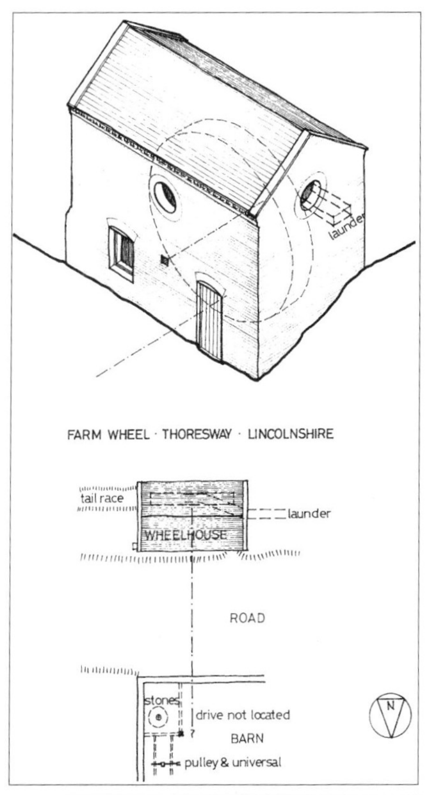

The layout of the wheelhouse and drive to the threshing mill at Thoresway, Lincolnshire.

forms part of a farm complex at Chollerton, Northumberland, and at Hoddam, Dumfries, a horse wheel superseded a windmill to drive a thresher.

In his *General View of the Agriculture of the County of Lincolnshire* of 1813, Arthur Young describes a mill built by Mr Holdgate at Thoresway, where an overshot waterwheel located in a brick wheelhouse drove a threshing machine in a barn on the opposite side of the road. Water was conducted a considerable distance 'with a true spirit of exertion', a reservoir having been formed on the side of a hill with the water carried 'in troughs upon trussle posts 20 ft. [6 metres] high' to the top of the wheel. An alternative was to site a waterwheel close to a convenient water supply and take the drive to the farm or barn. Such drives were usually taken off ring gears by pinions and shafts, but rope drives were also used. Many examples have been found in south-west England, particularly in Cornwall, where the influence of mining technology undoubtedly played a part in setting up drives that may be up to half a mile (0.8 km) in length. Some farm waterwheels are of considerable size and represent a high capital investment. On Lord Hatherton's estate at Teddesley, Staffordshire, an underground chamber was cut into solid rock to install an overshot wheel that was used to drive machinery (for threshing, chaff cutting, kibbling oats and barley for animal feed), millstones and circular saws. The original wheel was of

The remains of the overshot waterwheel and take-off from the pitwheel at Thoresway, Lincolnshire.

Overground drive from a breastshot waterwheel to a barn, for driving millstones and farm machinery at Patney, Wiltshire, in the 1950s.

timber and 30 feet (9.1 metres) in diameter but was replaced by a lightweight iron wheel some 8 feet (2.4 metres) larger in the 1840s. An additional benefit was that the wheel was driven by water drained from the surrounding land as part of a major improvement scheme.

While there are a number of eighteenth- and early-nineteenth-century illustrations of designs for farmsteads that include windmills, built examples are uncommon, probably because of the unreliability of wind and the high cost of maintenance compared with a horse gear or waterwheel. At West Blatchington, Sussex, there is a rare example of a smock mill built on top of a barn in the 1820s to drive millstones, a threshing machine and a chaff cutter. From the 1860s smaller *wind engines* were developed for driving agricultural machinery, such as the 'Unrivalled Self-Regulating Wind Engines' made by Bury & Pollard of Southwark, London. Their design was for a multi-sail windwheel winded by a fantail, supported on a timber frame above the roof of a barn, the drive being taken by light iron shafting and gearing to the millstones or barn machinery. Part of their advertising claimed that the machine could be repaired, if necessary, by the village blacksmith or carpenter and was 'free from all the risks incidental to steam'. Although steam power was advocated by agricultural improvers from the middle of the nineteenth century, water and wind power continued to be viable for small-scale intermittent use and, once set up, the use of machinery driven by a free source of power, and which could be maintained by local craftsmen, undoubtedly appealed to farmers.

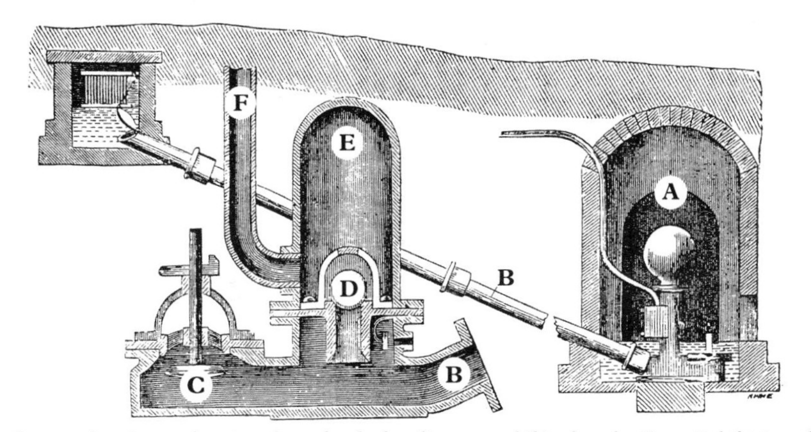

Layout drawing and section through a hydraulic ram, exhibited at the Great Exhibition of 1851 by Easton & Amos of Southwark, London. The layout drawing shows the ram in an underground chamber, A, fed by the drive pipe, B. The section through the ram shows the drive pipe, B, entering from the right, the waste valve, C, the delivery valve, D, at the bottom of the air vessel, E, and the delivery pipe, F.

Hydraulic rams

As well as obtaining rotary motion from moving or falling water, a number of hydraulic devices were invented to make use of water pressure. The first water-pressure engine in England was erected by William Westgarth for pumping water from a lead mine in Northumberland in 1765 and was favourably commented on by John Smeaton. Similar engines were used at Alport, Derbyshire, in the nineteenth century, and these economic devices received the attention of the mining engineer John Taylor and also of William Armstrong, who used hydraulic power to work dockyard cranes. Hydraulic rams are simple water-pumping devices that utilise a small fall in a stream to pump a fraction of that fall to a greater height. They became widely used because of their ability to deliver a small but regular amount of water, suitable for farm and village supply, with little need for supervision or maintenance. The hydraulic ram was invented in 1793 by the Frenchman Joseph Michel Montgolfier (1740–1810), who is remembered mainly for his hot-air balloon. Rams were made in Britain from the 1820s and, although they are simple devices, their detailed action is complex. The working principle is that water flows down a drive pipe into a valve box that contains a waste valve and a delivery valve. As the flowing water accelerates there comes a point when the hydraulic drag on the waste valve overcomes its weight and it closes, against gravity, with a characteristic thud. The sudden change in the flow leads to a rise in pressure that opens up the delivery valve and causes water to flow upwards into an air vessel and delivery pipe. Once the flow stops, the delivery valve closes and the waste valve opens, so allowing the cycle to repeat. There were a number of different models and makers, and rams can still be found around the countryside, often in remote pump houses, some of which are still in use.

From millstones to steel rollers

At Sibsey in Lincolnshire a new six-sailed brick tower mill was built in 1877 by the millwrights Saunderson of Louth to replace a post mill, and there are other similar instances elsewhere. New tower mills were also built at Stert, near Devizes, Wiltshire, in 1885 and at Much Hadham, Hertfordshire, in 1892–3, for example. What makes this even more remarkable is that this was the period when a revolution was taking place in the milling industry. The use of metal rollers for bruising and crushing grain, particularly oats and malt, has a history dating back to the seventeenth century, and in the late eighteenth century a pair of iron rolls became a standard addition to the machinery of oil mills. In the middle of the nineteenth century *roller mills* were experimented with as part of the flour-milling process, being used initially in combination with millstones. By the late 1870s new milling systems using chilled iron rolls to break and reduce the grain were being built in England, and in the last two decades of the nineteenth century roller-milling almost completely supplanted stone-milling in the production of flour. The new system of milling brought about a complete and radical change in technology, work organisation and location, with steam-driven mills that were in continuous production being sited not only at major inland centres but also at ports, to take advantage of grain imported from overseas, chiefly from North America. The country mills, powered by water and wind, could compete only as long as local demand continued. Some complete roller plants were built to be water-driven,

A fine milling complex at Stotfold, Bedfordshire, in 1980, with the weatherboarded watermill, steam-engine house and brick-built roller mill to the right. The watermill was severely damaged by fire in the early 1990s.

usually by turbines, as at Caudwell's Mill, Rowsley, Derbyshire, which was refitted in 1905. Many smaller mills, both water- and wind-powered, introduced some of the new technology, in the form of one or two pairs of rolls and improved sifting and dressing machinery, in an attempt to remain competitive with white-flour production. It is of interest to note that the mills of a number of the smaller independent millers who are still producing flour developed around traditional millstone mills, as at Timm's Mill, Goole, East Yorkshire, where the tower of the five-sailed windmill rented by Edward Timm in 1854, along with a steam mill close by, still stands.

Although windmills and watermills were still relatively inexpensive to run, many closed during the last two decades of the nineteenth century, and those that continued in work were often reduced to producing animal feeds. By 1900 many traditional mills also used auxiliary power, supplied by steam, gas or oil engines. The need for maintenance and the growing cost of repairs, particularly with windmills, constituted the main reasons for the smaller mills becoming uneconomic and being demolished. One notable exception, where a watermill was updated with millstone plant in the 1880s, is the House Mill, one of two tide mills at Bromley by Bow, London. The mill was rebuilt on an ancient site in 1776 with four waterwheels driving eight pairs of millstones. In 1886 a new cast-iron waterwheel was installed on the west side and two of the other waterwheels were improved by being partly rebuilt and fitted with curved metal floats. While the two wheels on the east side each continued to drive two pairs of stones by conventional spur gearing, new layshaft drives were put in on the west side and the two waterwheels on that side were geared to drive eight pairs of stones. The stones were set up on individual cast-iron hurst frames known as standards. This set-up represents one of the last major refurbishments of a large watermill using the best of nineteenth-century millstone technology and was probably carried out because the mill was used only for grinding maize for Nicholson's gin distillery, of which it was part. The economy of water power meant that the mill was both well sited and well suited to the continuing production of maize meal for distilling.

The Poncelet wheel

During the 1820s the initiative for improving water-powered prime movers passed to France. After the upheaval of the Napoleonic Wars there was a desire to catch up with industrialised Britain and, as there was less potential in France for steam power owing to a lack of coal, the development of water power was encouraged. Improved forms of low-head waterwheels and water turbines were designed and built, notably by Jean Victor Poncelet and Benoît Fourneyron.

The name 'Poncelet' is often rather casually applied to undershot and breastshot waterwheels with curved metal floats that survive in British mills. Such wheels should not be defined as anything other than 'improved undershot' unless they combine the distinctive features of the rational design for an undershot wheel that was published after a period of experiment by the French

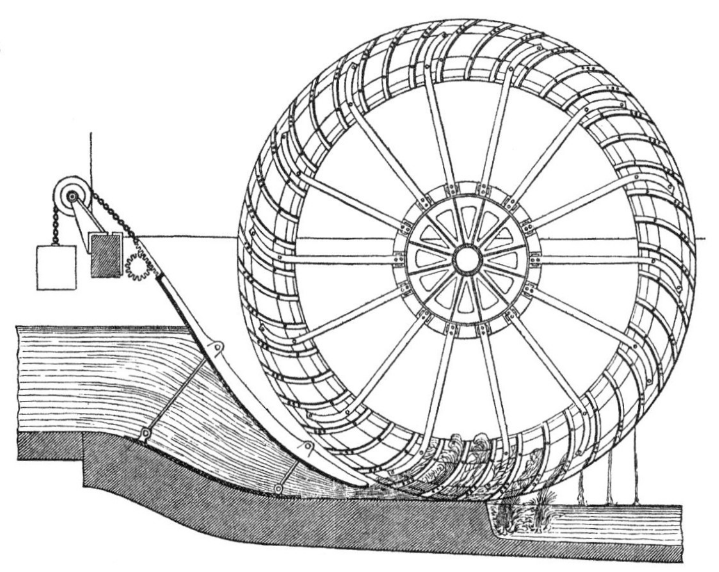

Poncelet's waterwheel design, from William Fairbairn's 'Treatise on Mills and Millwork', 1861. Note the distinctive features of closely spaced, curved, shrouded buckets without sole plates, a penstock that permits close control of water entry, and a fall in the wheelpit immediately below the wheel.

engineer General J. V. Poncelet in 1825. A note in the trade magazine *The Miller* for 6th June 1888 succinctly sums up the qualities of this particular design:

> The Poncelet water wheel owes its superiority to the construction of the buckets which are curved and form an angle with the inlet of the water so that the latter enters the bucket without shock, imposing its initial velocity to the buckets, mounting along them and acting by its weight on leaving them. The principle on which the water acts on the wheel is the same as that of the impulse or action turbine. The buckets have side boards [shrouds] so that no water can escape without having done its work. The water should be admitted as close to the wheel as possible. The Poncelet wheel should be entirely of iron.

It was further noted that such wheels could easily be adapted to locations where undershot wheels were working, although few Poncelet wheels have been identified in Britain and, where they were built, they were usually of particular note. One such wheel was installed at Exwick Mill, Devon, in the 1880s by the local engineer Alfred Bodley and received much attention in the milling and engineering press of the time. In his discussion of Poncelet's design in *Mills and Millwork*, William Fairbairn gives figures of fifty to sixty per cent efficiency, but notes that, while such wheels may be used with advantage on falls up to about 6 feet (1.8 metres) 'above this the low breast wheel is certainly more advantageous and costs less'. It was probably the difference in cost coupled with the high engineering requirement necessary to produce wheels to Poncelet's precise design that led many British millwrights and iron founders to develop improved, but technically inferior, undershot and low breastshot wheels with curved floats.

Water turbines

The two fundamental requirements for the perfect prime mover, which were identified during the eighteenth century, are that water must enter without shock and leave without velocity. These were addressed by Poncelet in designing his buckets on mathematical principles and putting a shelf in the race immediately below the wheel in order that water would fall away quickly and cleanly, but he also appreciated the limitations of his design, with water entering and leaving the wheel at about the same point. At the same time that he was working on his vertical wheel, Benoît Fourneyron was developing a horizontal waterwheel that could work completely submerged in water.

The history and development of water turbines is complicated by the fact that they are difficult to define precisely and should not be seen as anything other than rotating hydraulic motors that were formerly called waterwheels and were developed and improved over a period of time. Conventional waterwheels are basically slow-moving and generally of massive construction and highly stressed, and so in need of frequent maintenance and repair. There is also a limit to the head of water that can be effectively utilised, usually about 50 feet (15.2 metres). Water turbines occupy a smaller space, run at a comparatively higher speed, can work under a wide variety of heads and produce greater power than waterwheels. They can also work fully immersed in water. A well-built water turbine should have an effective working life of thirty to fifty years if well maintained and run with a clean water supply. The name 'turbine' was probably first used by the French engineer Claude Burdin, a contemporary of Fourneyron.

There are basically two types of turbine, described as *reaction* and *impulse.* In a reaction turbine, all the water passages are completely filled and the energy stored in the water at the inlet of the turbine is transferred to the wheel or runner as it passes through the machine. The machine acts under pressure and can be completely submerged. In an impulse turbine, water is directed on to the wheel from one or more jets, so the water is delivered at a velocity due to its head. The rotor loses power if the tailwater touches it.

The simplest form of reaction turbine was that which Dr Barker communicated to the natural philosopher J. T. Desaguliers in about 1740. Barker's mill, which was really only a philosophical toy until the mid nineteenth

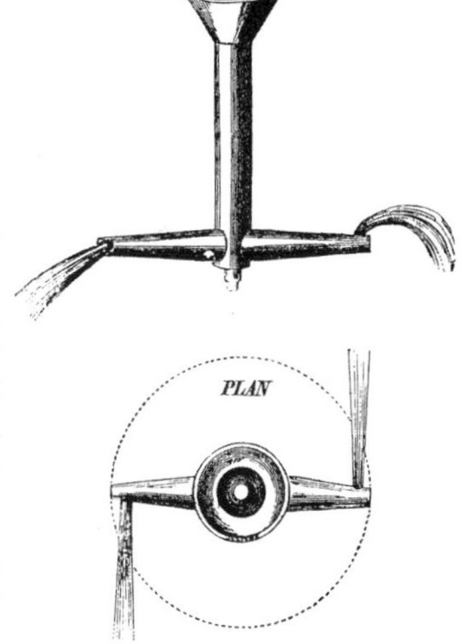

The principle of Barker's mill, a simple reaction turbine of about 1740, from Joseph Glynn's 'Power of Water', 1879.

Scotch turbine by David Cook & Company of Glasgow, 1857, at Sutton Poyntz Pumping Station, Dorset.

century, comprised a vertical tube with two horizontal arms projecting from it. Water introduced into the vertical tube spouted under pressure from holes in the ends of the arms, which turned the vertical tube by reaction to these jets. The principle is still used for lawn sprinklers and the turning arms on sewage filter tanks. The idea was developed by James Whitelaw of Glasgow, who replaced the straight arms of Barker's mill with an S-shape, a design which he patented in 1839 and which became known as the *Scotch mill.* The first practical example was built by Donald & Craig of Paisley for Mr Stirrat, to drive a powerful water press, and thereafter Whitelaw and Stirrat worked together on producing turbines. William Cubitt used one for hauling boats up an inclined plane on the Chard Canal in Somerset after 1842, and two built by David Cook & Company of Glasgow were installed at the Weymouth Waterworks Company's pumping station at Sutton Poyntz, Dorset, in 1856 and 1857. The 1857 turbine, which was capable of pumping 300,000 gallons (1.4 million litres) of water per day, survives. Two Whitelaw and Stirrat turbines were installed at Frocester Court, Gloucestershire, after 1859: a small one, which was dismantled about 1900, for driving dairy machinery, and a larger one, which was installed in a pit in the tithe barn, for driving barn machinery. It is doubtful that many Scotch mills were produced after about 1869.

The major advance in turbine design was that made by Benoît Fourneyron (1802–67), who developed a compact, fast-running and high-power generating waterwheel in 1827, for which he was subsequently awarded a prize. Fourneyron, a good practical engineer with experience in a wide range of engineering and industrial work, produced a machine in which water was introduced into the centre, from above or below, through a fixed circle of curved vanes to turn a horizontal runner that drove a vertical shaft, now defined as an outward-flow horizontal reaction turbine. Between 1832 and his death Fourneyron designed and installed over a hundred turbines in France and elsewhere in Europe which worked with notable success. There was considerably more interest in using water turbines in the textile industry of Northern Ireland, where there were useful streams, a high rainfall and a lack of coal,

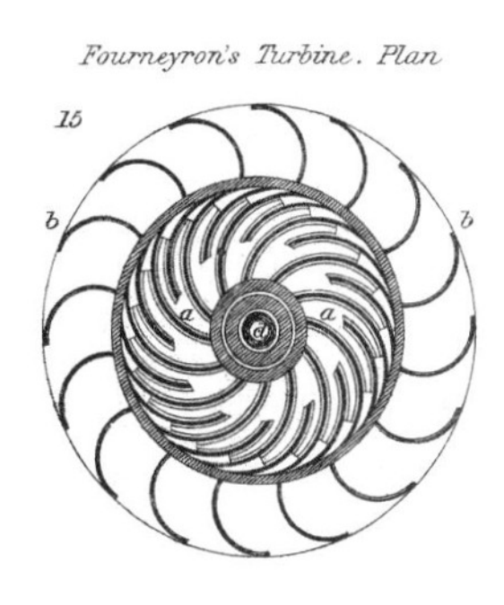

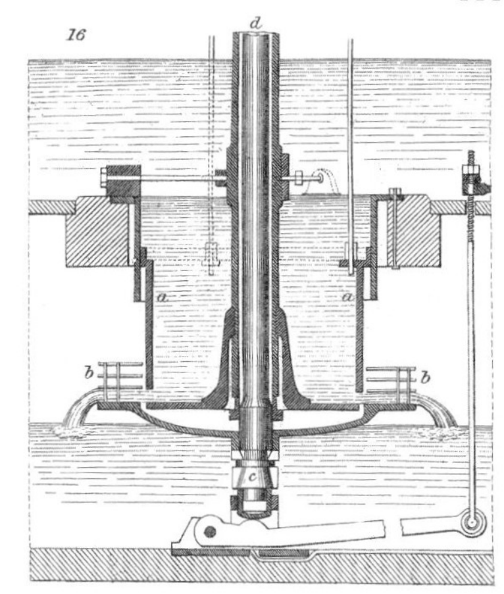

Plan and section of Fourneyron's turbine, from a nineteenth-century encyclopaedia. Water enters the centre of the machine and is directed by the guide vanes of the stator, aa, outwards through the rotor, bb.

than in England, where steam power dominated. In about 1850 William Cullen, a millwright from Armagh, visited France to find out about Fourneyron's turbines and, although the French engineer was uncooperative, Cullen managed to see some examples and plans, and on his return to Northern Ireland was able to build a model. Cullen did not have the resources to manufacture turbines himself, however, and he developed Fourneyron's design with Robert MacAdam of the Soho Foundry, Belfast. By 1860 the MacAdam Brothers had installed a number of turbines in spinning and flax-scutching mills and in about 1870 they supplied a turbine for Catteshall Mill, on the river Wey at

The rotor and stator of the MacAdam turbine salvaged from Catteshall Mill, Godalming, Surrey, in the 1980s.

An Armfield double 'British Empire' turbine: a catalogue illustration of about 1900, showing the control mechanism for the two gates that admitted water into the turbine casing. The outlets of the two rotors can be seen to the top and bottom of the casing on the right.

Godalming, Surrey. The mill occupied a site that had been used from the eleventh century for corn and malt milling, fulling, papermaking, tanning, and engineering and foundry work. From the mid nineteenth century power was provided by steam, and the water turbine, working on a head of about 6 feet 6 inches (2.0 metres), replaced a waterwheel. The MacAdam turbine was rescued for conservation and display in the early 1980s.

Another form of reaction turbine developed from Fourneyron's design, but in which the water flowed inward through adjustable guide vanes, was patented by James Thompson (1822–92) of Belfast in 1850. Known as the 'Vortex' turbine, from the action of the water flow, this proved to be a very adaptable machine, capable of working on a wide range of heads from 3 to 380 feet (1 to 125 metres). The first working machine was built in Glasgow and installed in a beetling mill in County Antrim in 1852. While no British-made turbines were shown at the Great Exhibition of 1851, four manufacturers were represented at an exhibition in 1862, including Williamson Brothers of the Canal Ironworks, Kendal, Cumbria, who built their first turbine in 1856 under licence from Thompson. It was used for driving farm machinery and produced 5 horsepower (3.75 kW) on a head of about 35 feet (10 metres). In 1858 they installed a turbine at Stott Park Bobbin Mill, Cumbria, and in 1865 began exporting, their output averaging about seventeen machines a year up to 1881. In that year the company was bought out by Gilbert Gilkes (1845–1924), who was born in Dublin and had trained as an engineer in Middlesbrough. Up to 1900, Gilkes's average output was fifty-two turbines a year, of substantially higher power than the Williamsons'.

Reaction turbines were also developed in the United States, one of the most successful designs being that of James Francis (1815–92). Sometime after 1850 Francis, who worked in Massachusetts, developed an inward-flow turbine with movable guide vanes that controlled the water entry more efficiently than Fourneyron's original design. The machine could be adapted for either vertical or horizontal drives and worked on heads up to about 175 feet (50 metres). Many American manufacturers concentrated on Francis's designs, and in the last quarter of the nineteenth century a large number of their machines were imported into Britain. The main engineering and millwrighting

A Pelton wheel, with supply pipe and nozzles to the left, latterly used for driving a saw at East Portlemouth, Devon.

company involved with the installation of turbines in the south of England was Armfield of Ringwood, Hampshire. The company was started by William Munden, who took on Joseph J. Armfield (1852–1938), a young millwright, in about 1875. Armfield took over the business in 1888 and by 1900 the firm had built ninety-eight turbines, most of which were installed to replace waterwheels. Armfield's designs, known as the 'British Empire' and 'River Patent' turbines, were based on contemporary American practice. Many examples survive, some in working order, such as the 45 inch (1.1 metre) British Empire turbine installed in 1904 at Sturminster Newton Mill on the river Stour, Dorset. At Boar Mill, Corfe Castle, Dorset, a 10 inch (0.25 metre) machine was put in for the miller and baker Charles Battrick in 1901. The work involved removing an overshot waterwheel and the spur gearing that drove two pairs of millstones; the Armfield turbine, including shafting and belt drives, cost £90.

The impulse wheel was also developed in the United States, in the Californian gold-mining area. An early, inefficient form was the 'hurdy-gurdy', simply a metal pulley with a series of plates fixed to its periphery to form paddles, turned by water being directed on to it through a pipe and jet nozzle. In 1880 an improved design in which double cup-shaped buckets replaced the flat plates was patented by Lester A. Pelton. While 'Pelton wheels' are highly efficient, in practice it was found that there was a limit to the size of jet relative to that of wheel, which restricted the overall size and thus the power that could be produced. The first Pelton wheels to be built in England were probably manufactured by Charles Hett of Brigg, Lincolnshire, in about 1890. Hett also introduced several well-known American turbines into England, some of which were built by Gilkes, who eventually took over his business in 1895. There were also devices described as 'water motors', such as that patented by John, Henry and William Chidgey of Watchet, Somerset, in 1894, which was a form of impulse turbine enclosed in an air-tight case. There are numerous applications of water power from this period and some of the surviving examples still need to be researched and recorded.

In 1928 Gilkes bought out the water-power side of James Gordon & Company of London and the firm became Gilbert Gilkes & Gordon. They subsequently took over several other concerns, including Günther & Sons of Oldham, Lancashire, in 1933. Günther was of German extraction and his turbines were based on continental designs. He manufactured turbines of the outward-flow type, including those developed in Europe by Girard and Jonval, but neither of these designs was widely adopted. All the specific designs for water turbines were developed before 1900, with the exception of the 'Turgo', a high-speed impulse wheel patented by Eric Crewdson of Gilbert Gilkes & Company in 1920, and the higher-speed propeller-type reaction turbines, such as that developed by Victor Kaplan between 1910 and 1924.

Hydroelectricity

The world's first hydroelectric installation was for lighting the home of Sir William Armstrong at Cragside, near Rothbury, Northumberland, in 1879–80, using incandescent light bulbs, which had been developed by Joseph Swan in Newcastle upon Tyne. Power was supplied by a turbine (number 428) made by Williamson Brothers, which developed 9 horsepower (6.75 kW) on a head of 29 feet (8.8 metres). There was a rapid growth of electric power supply during the last two decades of the nineteenth century, but hydroelectric generation was never particularly important in England because of the small potential relative to other countries and also because adequate water power was only widely available in the wrong places, remote from towns. The first public-supply installation was set up at Westbrook Mill on the river Wey at Godalming, Surrey, in 1881, using a breastshot waterwheel to drive a dynamo. The wheel proved inadequate and a second was coupled to it, but the venture was short-lived and was abandoned in favour of gas lighting in 1884. The first town to generate electricity for public supply in Scotland was Greenock in 1885, using dynamos rope-driven from a 40 horsepower (30 kW) Girard-type turbine made by Günther of Oldham. Again the scheme was short-lived, being abandoned in 1887. The earliest successful schemes were both offshoots of private ventures. At Wickwar, Gloucestershire, a 36 foot (10.9 metre) diameter overshot wheel that drove a generator at a brewery supplied surplus electricity for lighting part of the town in 1888, and at Okehampton, Devon, a similar arrangement existed using power from a turbine installed to drive the machinery of a sawmill owned by Henry Geen. A steam engine was added in 1896, and the generating plant was taken over by the West Devon Electric Supply Company in 1930, finally closing in 1937, when a new hydroelectric station was opened at Mary Tavy. Henry Geen's brother Charles set up an enterprising system of public lighting at Lynmouth, north Devon, in 1890, using water from the East Lyn river to drive a 'Little Giant' turbine built by Charles Hett of Brigg. By 1895 demand had outgrown supply and the water supply was a problem because of dry weather and also its retention by a flour miller upstream. A high-level reservoir was therefore constructed to which water was pumped during the day ('off peak') and, when the demand for electricity increased in the evening, the water was released to drive two Pelton wheels. This is the first scheme that is known to have made use of a pumped storage system in connection with electricity generation. Water power was supplemented by steam and subsequently by oil engines from 1923, and the system continued until it was destroyed by the floods of August 1952.

There were nine hydro plants producing electricity for public supply in England and Scotland before 1900, of which the largest was at Worcester, which was capable of generating more than an average steam-powered station of the same period. Three of the early schemes were in Devon, which still produces more hydroelectricity than any other English county, with generating stations at Morwellham, using water from John Taylor's canal, and at Mary Tavy. Numerous small hydro-power schemes have been built during the twentieth century, using both conventional waterwheels and turbines; many of these survive and still produce electricity, putting surplus into the National Grid. In Devon an abbey and a castle both still use water power to supply electricity. Buckfast Abbey, rebuilt on the ruins of a medieval Cistercian foundation by Benedictine monks in the early years of the twentieth century, uses turbines and a waterwheel installed in 1997 to produce hydroelectricity. At Castle Drogo, Drewsteignton, a small generating plant was set up in 1928 to supply electricity to the castle then being built for Julius Drewe. Two Francis-type Gilkes turbines of 27 and 55 horsepower (20 and 41 kW) were installed to take account of the seasonal fluctuations of power requirements and water levels in the river Teign. The turbines and generating set are located in a reinforced concrete 'cottage orné', built as part of the scheme designed by Sir Edwin Lutyens. At Aberdulais Falls, near Neath in South Wales, an industrial site used for copper manufacture, corn milling and, more recently, tinplate production now makes use of the Dulais river and falls to produce electricity

Modern overshot waterwheel for electricity generation at Aberdulais Falls, Neath, South Wales.

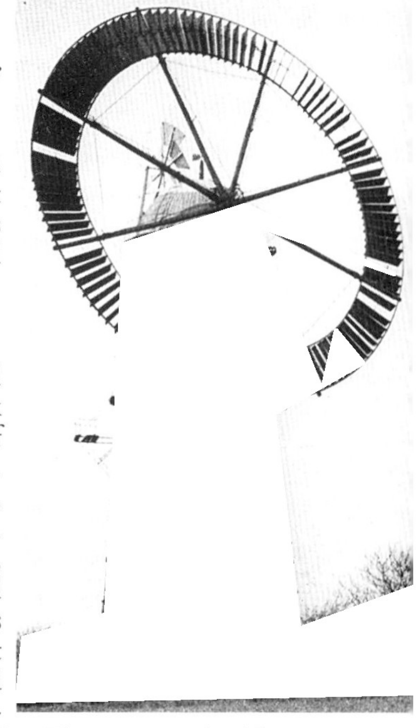

Annular-sailed windmill at Haverhill, Suffolk, in the early 1930s.

with a Swedish axial flow turbine and a modern waterwheel. The waterwheel, which is overshot and 27 feet (8.2 metres) in diameter, produces about 34 horsepower (25 kW) and was installed in 1992.

Wind engines

In the nineteenth century the conventional form of windmill sail was a timber framework covered with canvas or divided into a series of bays that were filled with movable shutters. Windmills commonly worked with four sails, but five-, six- and eight-sailers were also built. In 1866 Henry Chopping, a miller at Roxwell, Essex, put up what was described as a 'wheel sail' on his post mill, after its conventional sails were blown off in a gale. The wheel was 52 feet (15.8 metres) in diameter with timber-framed canvas-covered shutters about 7 feet (2.1 metres) in length. This *annular sail* was stated by Chopping to be his own invention and to be superior to conventional sails. He further claimed that it was capable of driving four pairs of stones; if correct, this is the only record of that number in an Essex post mill. Similar annular sails were erected on a smock mill at Boxford and a tower mill at Haverhill, both in Suffolk, the latter standing into the 1940s. In 1868 Chopping took out a patent with F. Warner for an annular-sailed wind pump in which the ring of shutters was divided into three or more parts, each section acting independently to feather or turn out of wind.

Annular-sailed wind pumps built predominantly of steel were developed in America in the 1870s and used in Britain from the 1880s. There were a

Wind engine with tank for holding pumped water, on the chalk down at Homington, Wiltshire.

Three-megawatt wind generator, the largest in the British Isles, on Burgar Hill, Orkney.

number of manufacturers, most of whom were already established in agricultural engineering and water-supply works. Thomas & Son of Worcester, established in 1822, who made the 'Climax' wind engine, won the second prize in the Royal Agricultural Society of England's tests on wind pumps that were carried out in 1903. Twenty-two different models by seventeen manufacturers were assessed, the winner being a Canadian machine. Thomas & Son produced some thirty thousand wind pumps before their business closed in the early 1970s. Other manufacturers include John Wallis Titt of the Woodcock Ironworks, Warminster, Wiltshire, who started building a range of wind engines in the 1880s, mainly for pumping but also for driving farm machinery and for milling. At Crux Easton, Hampshire, a 'Simplex' wind engine with sails 20 feet (6 metres) in diameter on a tower 30 feet (9.1 metres) high, erected in 1894, drove pumps and millstones and was capable of grinding 'eight to ten sacks of corn per day' (over 1 tonne). In 1895 John Wallis Titt also set up a Simplex engine alongside a horse engine at Hinton Charterhouse, Somerset, which was claimed to be the first village to have a pumped water supply. The work was undertaken for the Bath Rural District Council, whose engineer was pleased with the results, particularly as there was 'no expense beyond oiling'.

Wind engines are well suited to pumping as annular sails produce high torque and low speed, which is good for starting pumps working against a static head of water. Pumping water little and often made them economical and they continued in widespread use until the Water Resources Act of 1963 was implemented in April 1969, when charges levied on water extraction had a detrimental effect on many users of natural power. The National Association of Water Power Users was formed to combat the restrictive legislation and eventually succeeded in making small-scale hydro-power use a continuing worthwhile proposition.

Wind engines were used for electricity generation from the 1890s, but the irregularity of wind caused problems. The engineer and wind power enthusiast Edward Lancaster Burne undertook a series of experiments in wind generation in the early years of the twentieth century, building small-scale semi-conventional windmills with self-regulating sails and fantails to drive dynamos and electric plant. The last windmill to be built in Kent was at St Margaret's Bay in 1928. It was constructed by Holman Brothers, the Canterbury millwrights, and, while its external appearance is that of a conventional Kentish smock mill, it was built for electricity generation. In the second half of the twentieth century wind turbines were developed for power generation. A number of individual machines of varying designs were put up in the 1980s, the largest of which is the 3 megawatt wind generator at Burgar Hill, on Mainland, Orkney, built in 1987. In the 1990s a number of wind farms have been developed, particularly in Cornwall, northern England and Wales, with up to twenty-four wind turbines grouped together.

The watermill at Houghton, on the river Ouse in Cambridgeshire, viewed from downstream.

PRESERVATION

Before 1929 little effort had been made to protect watermills and windmills. A small number were preserved either as landmarks, such as the tower mill at Bidston, Cheshire, restored by Hudson, the soap manufacturer, in 1894, or as antiquarian curiosities, as with the horizontal-wheeled mill at Troswick, Shetland, restored by Gilbert Goudie sometime before 1904. Many mills were still at work, the most satisfactory way to ensure their survival at least in the short term, and there were still a number of millwrighting firms able to maintain and repair them.

In June 1929 the Society for the Protection of Ancient Buildings, founded by William Morris in 1877, was considering a new campaign to save and protect barns when it was asked to make a statement about the future of windmills in England. As a result of publicity in the national press, the Society's office was inundated with correspondence about and illustrations of mills. By 1931 the pressure was such that a separate Windmill Committee was formed, with the engineer and windmill enthusiast Rex Wailes (1901–86) as its technical adviser. Books and surveys were promoted and published, certificates awarded to those who showed 'zeal' in the maintenance, repair and upkeep of windmills, and money raised by appeal enabled some mills to be kept standing and others to continue working. In 1946 the role of the Windmill Section was extended to include watermills, and further county surveys and publications resulted. A survey of Lincolnshire windmills by Rex Wailes compared their condition in 1951 with that in 1923. Of some 215 post and tower mills recorded, 103 were working in 1923; in 1951 only six were still at work and a further two were preserved. A more recent survey published by Peter Dolman in 1986 lists a hundred and thirty remains of windmills in the old county of Lincolnshire, with eight mills working or preserved, a figure that increased to ten during the following decade. In 1951 the Ministry of Public Buildings and Works, Ancient Monuments Department (now English Heritage) received reports by Rex Wailes on twelve windmills and took into care the post mill at Saxtead Green, Suffolk, and Berney Arms tower mill, Norfolk. Trader Mill, the six-sailed tower mill at Sibsey, Lincolnshire, which was still working at that time, was taken into care shortly after. The china-stone grinding mills at Tregargus, near St Austell, Cornwall, were scheduled as an ancient monument as the result of a threat of demolition in 1968 and, while several other sites are similarly scheduled, most mills are now protected by listed building legislation, which includes the working parts as well as the buildings. However, long-term protection and preservation still rely ultimately on sympathetic ownership.

By 1960 there were few commercially operating watermills or windmills, and many mills that are now looked after and opened to the public have survived only because of the work of enthusiasts who carried out emergency holding repairs in the 1960s and 1970s. Norfolk County Council set up a

Restored waterwheel and launder at the Killhope Lead Mining Centre, County Durham.

fund for the restoration of windmills in 1960 and, as a result, the Norfolk Windmills Trust was formed in 1963, initially taking on eighteen windmills and wind pumps in a major restoration and maintenance programme. There is still a small number of working millwrights, some of whom started as volunteers and enthusiasts, but only Thompsons of Alford, Lincolnshire, can claim a continuity of traditional practice dating back to the nineteenth century. Essex County Council is the only authority to employ a full-time millwright to look after its mills, a job which has been done by Vincent Pargeter since the post was created in 1975.

A number of mills open to the public are in the care of official bodies such as English Heritage, the National Trust and regional or local authorities, but many are still in private hands and are looked after by individuals, small groups and private trusts, often with great success. A number of regional and local mill groups have also been formed, usually with a small but enthusiastic membership that works on recording, repairing and demonstrating mills to visitors. In the 1970s and 1980s a growing interest in wholesome foods and natural sources of power led to a revival of traditional milling, with an increase in the number of mills being restored to working order. In 1987 the Traditional Corn Millers Guild was formed to stimulate the interest in using water and wind power to produce traditional stoneground products, and in 1998 the Guild had thirty member mills spread throughout England, Wales, Scotland and Ireland.

The form and direction of mill preservation have changed considerably in the last decades of the twentieth century, but the work is far from finished. Even with more enlightened attitudes, tighter legislation and better funding, there is still a great need for care and understanding towards those windmills and watermills that have survived. As with any other working engines, skilled supervision and regular maintenance are essential, even after thorough repair and restoration. Historical and archaeological research, as well as fieldwork and recording, are vital to increase understanding of this important part of Britain's history. Mills may once have been common and widespread, but they are no less important in historical terms than churches, castles and manor houses, with which many share their origins. Their modest scale has always made them vulnerable to conversion and, because they are machines, to adaptation to other uses, which is an important part of their history. The study of water- and wind-powered machinery has great educational value in terms of learning about the development of mechanical engineering, the use of materials, the processing of foodstuffs and the production of essential goods, as well as promoting a better understanding of the value of natural, renewable sources of power.

House Mill, Bromley by Bow, London. The mill stands on the lower, tidal reaches of the river Lea and contains four waterwheels, formerly driving twelve pairs of millstones. The mill house, to the right, and the mill were restored in the 1990s.

GLOSSARY

Annular sail: a circular form of sail with a single ring of radial shutters or blades.
Bark mill: used for grinding the bark of trees, usually oak, for the preparation of tannin for making leather.
Bearing: the static part of a machine in which a journal runs.
Beetling mill: a mill in which linen cloth was pounded to produce a sheen.
Bloomery: a water-powered forge in which blooms (or bars) of wrought iron were produced direct from ore by smelting and hammering.
Blowing mill: a mill using water-powered bellows to smelt metallic ore.
Boat mill: see *Floating mill.*
Bob: a counterweight on a rocking beam, used to balance the weight of pump or flat rods driven by a crank from a waterwheel. Its name derives from its up and down movement.
Bobbin: in textile mills, a spool on to which spun thread is wound.
Brakewheel: the primary gear mounted on the windshaft in windmills, on which the brake acts.
Breastshot wheel: a vertical waterwheel where the water enters at about the level of the wheelshaft, driven by both the impulse and the weight of the water.
Buddle: a structure used in mining in which water is used to separate metallic ore from lighter particles – a form of settlement tank.
Cam: a projection on a wheel or shaft to transmit movement to another part of the machinery.
Clasp arm: a form of construction used for waterwheels and gear wheels where two pairs of arms form a square at the centre that boxes the shaft on to which the wheel is fixed.
Coalmill: a waterwheel-driven pump used for raising water from a coal mine.
Cog: an individual timber tooth inserted into a gearwheel.
Common sail: the earliest form of windmill sail where cloth is spread over a lattice framework.
Compass arm: a form of construction in which the arms of a waterwheel or gear are mortised through the shaft.
Crazing mill: a waterwheel-driven mill in which millstones are used to grind tin ore.
Crosstrees: the main horizontal timbers of the trestle of a post mill, from which the quarterbars rise to support the post.
Crown wheel: a horizontal-face gear, with its cogs or teeth usually projecting upwards, from which drives are taken by pinions and layshafts.
Dendrochronology: the scientific dating of timber by measuring the annual growth rings and comparing them with established chronologies.
Double mill: a mill that contains two sets of machinery or millstones, often driven by separate waterwheels. (Also, **treble mill:** three sets.)
Dressing: the art of preparing the working faces of millstones for grinding. Also used for sieving meal to make a finer flour.
Earth-fast: a building system in which the feet of the load-bearing members, such as timber posts, are set firmly into the ground in post pits or trenches.

Edge runner stones: a pair of vertically mounted stones that rotate on a fixed horizontal bedstone, used for crushing rather than grinding.
Eye: the hole through the centre of a millstone.
Fantail: a small wind wheel set at right angles to the sails of a windmill to turn the mill automatically into the wind.
Finery: a forge in which pig iron was converted into wrought iron.
Flat rods: horizontal timber or iron rods linking a crank driven by a waterwheel to the pumping machinery, which may be some distance from the waterwheel.
Floating or boat mill: a corn mill mounted on a barge or pontoon, driven by a waterwheel turned by the river on which the mill floats.
Fulling mill: a mill in which woven cloth is scoured and beaten to felt the fibres together. Also known as a 'tucking' or 'walk' mill.
Governor: a device for controlling the gap between millstones; also for regulating the speed of a rotating waterwheel.
Grindstone: a single, vertically mounted rotating stone used for sharpening tools.
Harps: a pattern of furrows laid out in triangular segments on a millstone.
Head and tail: the arrangement of two gearwheels mounted one behind the other on the windshaft of a post mill, from which drives to millstones are taken.
Hollow post mill: a post mill in which the drive is taken down to the base of the mill by a vertical shaft passing through the hollowed centre of the post.
Horizontal wheel: a waterwheel that rotates in a horizontal plane.
Horizontal windmill: a type of windmill in which the sails rotate in a horizontal plane.
Horse mill: a corn mill or machinery powered by a horse walking in a circle.
Hurst: the sturdy timber frame that supports the millstones in a corn mill or the hammers in a forge mill.
Impulse turbine: a form of water turbine in which jets of water are directed on to a rotor.
Incorporating mill: a mill in which the ingredients of gunpowder are mixed using edge runner stones.
Journal: circular part of a shaft, usually of metal, which runs in a bearing.
Lade: a Scottish term for a man-made millstream. (See *Leat.*)
Lantern pinion: a driven gear formed of two discs with staves between which serve as cogs.
Launder: a trough, usually of timber, that leads water on to a waterwheel.
Layshaft drive: a gearing layout in which the drive is transmitted by horizontal shafting and face or bevel gearing.
Leat: a man-made stream that brings water to a waterwheel or mill, called *lade* in Scotland and *goit* in Yorkshire.
Millstone: one of a pair of usually horizontal stones for grinding corn.
Millwright: traditionally, someone who builds and maintains mills.
Naves: the iron centres fixed to a wheelshaft from which the arms radiate.
Overdrift: machinery, particularly millstones, driven from above.
Overshot wheel: a waterwheel driven by water entering at the top and turning it by the weight of the water in its buckets.
Patent sail: a form of remotely regulated shutter sail patented in 1807.
Penstock: a sluice or hatch that regulates the flow of water on to a waterwheel or

turbine.

Pinion: the smaller wheel of two wheels in gear, and driven by the larger wheel. Sometimes referred to as a *nut*.

Pitchback wheel: a waterwheel in which water enters at the top but turns the wheel backwards, in the opposite direction to an overshot wheel.

Pitwheel: the primary driven gear in a watermill, usually fixed to the wheelshaft, with its lower half turning in a pit.

Poll end/canister: the outer end of a windshaft, to which the sails are attached.

Post mill: a timber-framed mill of which the body, containing the machinery and carrying the sails, rotates about the head of a massive vertical post.

Quarterbars: the raking struts rising from the crosstrees that support the post of a post mill.

Quern : a pair of small diameter millstones, turned by hand, usually for grinding grain.

Race: a channel bringing water to or from a millwheel.

Reaction turbine: a form of water turbine in which all the water passages are filled and the rotor is turned by the energy stored in the water as it passes through the machine.

Returning engine: a steam engine used to pump water on to a waterwheel.

Roller mill: a machine with cylindrical rollers for crushing grain or other raw materials. Also a type of mill developed during the nineteenth century in which a series of rolls in combination with sieves is used to produce fine flour.

Roller reefing sail: a form of remote-controlled shuttered sail patented in 1789.

Runner: the moving stone of a pair of millstones. Also a block of stone used in a grinding pan for reducing flints etc for making pottery glazes.

Rynd: an iron fitting on which the upper, moving millstone is hung.

Sail bars: the short lateral bars of a windmill sail.

Scoop wheel: a driven wheel used to raise water in land drainage.

Scotch mill: a reaction turbine with S-shaped arms patented in 1839.

Scouring: a term used to describe the finishing process in needle making.

Scutching mill: a water- or wind-powered mill for processing flax, in which the plant fibres are broken and beaten to remove the fibrous material which is made into linen.

Shuttered sail: a form of windmill sail which is divided into a series of bays filled with movable shutters.

Sliding hatch: a form of waterwheel penstock in which the gate is depressed so that water is fed on to a wheel over its top, thus utilising a better head.

Smock mill: a timber-framed tower windmill, in England usually clad with horizontal or vertical timber boarding.

Spider: the cranks at the centre of a set of patent sails which link the shutters to the striking rod.

Spindle: a small-diameter shaft, usually of iron.

Spring sail: the earliest form of shuttered sail, in which the shutters were held closed by tensioned springs.

Spurwheel drive: a gearing form in which a number of drives can be taken off the periphery of a spur gear. In a corn mill the spurwheel is usually horizontal and a number of pairs of millstones can be grouped around a central shaft.

Stampers: vertical timbers, sometimes shod with iron, raised by cams and used to break up or press raw materials.

Stamping mill: a mill in which stamps or hammers are used, usually for breaking up metallic ore.

Start and awe wheel: Scottish term for an undershot or breastshot waterwheel that has floats (awes) fixed to starts projecting from timber or iron rings.

Stocks: wooden hammers in a fulling mill for beating cloth to scour it.

Striking rod: a rod passing through the centre of a windshaft, which connects the shutters of patent sails to the striking gear, by which the shutters are remotely controlled.

Suspension wheel: a form of iron waterwheel in which heavy timber or iron arms are replaced by lightweight iron rods and cross braces that hold the structure in tension.

Tailpole: a beam at the bottom of a post-mill body or extending from the cap of a tower mill used to turn the mill to the wind.

Tailwheel: in a post mill, a gear wheel mounted towards the inner end of the windshaft that drives millstones located in the tail of the mill.

Tenter frame: a timber framework with rows of small metal hooks on which cloth is spread to dry and shrink after fulling or dyeing.

Threshing machine: a farm machine used for separating grain from straw and chaff after harvesting.

Throwing mill: a watermill in which raw silk is wound and spun.

Tide mill: a watermill worked by salt water that has been impounded at high tide and is released on to a waterwheel or wheels as the tide falls.

Tower mill: a windmill comprising a masonry tower containing the machinery and a rotating cap at the top carrying the windshaft and sails.

Trestle: the substructure of a post mill.

Trundle: an ancient term for a small driven gear, a lantern pinion.

Tucking: a West Country term for fulling.

Underdrift: machinery, particularly millstones, driven from below.

Undershot wheel: a waterwheel driven by the impulse of water striking the floats at or near the bottom of the wheel.

Vertical waterwheel: a waterwheel that rotates in a vertical plane.

Wallower: the first gear driven by the pitwheel in a watermill and the brakewheel in a windmill.

Water turbine: a developed form of waterwheel that can be fully immersed in water and work more efficiently, providing more power under a variety of heads.

Weather: the twist of a windmill sail to the plane of rotation, necessary to transform the wind's energy into motive power.

Wheelshaft: the main horizontal driveshaft in a watermill, on which a waterwheel is mounted.

Wind engine: a form of annular-sailed windmill, usually mounted on a skeleton-type tower.

Windshaft: the main driveshaft in a windmill that carries the sails at its outer end and is turned by them.

FURTHER READING

There is a great deal of published information concerned with the history and development of water and wind power and, specifically, of watermills and windmills. Some contemporary sources are referred to in the text and the following selection is given to amplify the information set out in this book.

For the origins of water and wind power:

Lewis, M.J.T. *Millstone and Hammer, the Origins of Water Power.* University of Hull, 1997.

Lewis, M.J.T. 'The Greeks and the Early Windmill', *History of Technology*, 15 (1993), 141–89.

Smith, N.A.F. 'The Origins of Water Power: a Problem of Evidence and Expectations', *Transactions of the Newcomen Society*, 55 (1983–4), 67–84.

Spain, R.J. 'Romano-British Watermills', *Archaeologia Cantiana*, 100 (1984), 101–28.

Information about specific medieval sites that have been identified by archaeology appears in the annual volumes of *Medieval Archaeology.* See also:

Astill, G. *A Medieval Industrial Complex and Its Landscape.* CBA (Council for British Archaeology) Research Report 92, 1993.

Bond, C.J. *Medieval Windmills in South-Western England.* SPAB, 1995.

Crossley, D.W. (editor). *Medieval Industry.* CBA Research Report 40, 1981.

Holt, Richard. *The Mills of Medieval England.* Blackwell, 1988.

Langdon, John. 'The Birth and Demise of a Medieval Windmill', *History of Technology*, 14 (1992), 54–76.

Rahtz, Philip, and Meeson, Robert. *An Anglo-Saxon Watermill at Tamworth.* CBA Research Report 83, 1992.

Rynne, Colin. *Technological Change in Anglo-Norman Munster.* Barryscourt Trust, Cork, 1998.

For the post-medieval period the most useful discussion of the archaeological evidence of power and a variety of uses is given in:

Crossley, David. *Post-medieval Archaeology.* Leicester University Press, 1990.

Mills and industrial sites are also reported in the annual volumes of *Post-medieval Archaeology.*

For the industrial period there is an abundance of source material including the local and regional industrial archaeological series published by David & Charles from the 1960s and Batsford from the 1970s. Also the many papers and articles in the annual volumes of the *Transactions of the Newcomen Society*, particularly the county windmill surveys by Rex Wailes, and *Industrial Archaeology Review.*

Watermills have not received the same level of coverage as windmills, the most reliable general books being:

Hills, Richard L. *Power from Wind*. Cambridge University Press, 1994.
Reynolds, John. *Windmills and Watermills*. Evelyn, 1970.
Reynolds, Terry S. *Stronger than a Hundred Men, a History of the Vertical Waterwheel*. Johns Hopkins University Press, 1983.
Wailes, Rex. *The English Windmill*. Routledge, 1971.

There are numerous regional and local studies and gazetteers of varying quality, but all are of value when seeking information. A good range of different approaches can be found in the following:

Apling, Harry. *Norfolk Corn Windmills*. The Norfolk Windmills Trust, 1984.
Benham, H. *Some Essex Watermills*. Essex County Newspapers, 1976.
Dolman, Peter. *Lincolnshire Windmills*. Lincolnshire County Council, 1986.
Farries, Kenneth. *Essex Windmills, Millers and Millwrights* (five volumes). Charles Skilton, 1981–8.
Foreman, Wilfred. *Oxfordshire Mills*. Phillimore, 1983.
Gregory, Roy. *East Yorkshire Windmills*. Charles Skilton, 1985.
Gribbon, H.D. *The History of Water Power in Ulster*. David & Charles, 1969.
Guise, Barry, and Lees, George. *Windmills of Anglesey*. Attic Books, 1992.
Shaw, John. *Water Power in Scotland*. John Donald, 1984.
Somervell, John. *Water-power Mills of South Westmorland*. Wilson, 1930.
Stidder, D., and Smith, C. *Watermills of Sussex*, 1. Baron, 1997.
Watts, Martin. *Wiltshire Windmills*. Wiltshire County Council, 1980.

For specific uses of water and wind power, there are a number of useful books and articles, starting with the classic:

Bennett, Richard, and Elton, John. *History of Corn Milling* (four volumes). Simpkin Marshall & Co, 1898–1904.
Callandine, Anthony, and Fricker, Jean. *East Cheshire Textile Mills*. RCHME, 1993.
Cleere, H., and Crossley, D. *The Iron Industry of the Weald*. Leicester University Press, 1985.
Day, Joan. *Bristol Brass: A History of the Industry*. David & Charles, 1973.
Ford, Trevor D., and Willies, Lynn (editors). *Mining Before Powder*. Peak District Mines Historical Society Bulletin 12, 1994.
Giles, Colum, and Goodall, Ian H. *Yorkshire Textile Mills 1770–1930*. HMSO, 1992.
Ingle, George. *Yorkshire Cotton*. Carnegie, 1997.
McCutcheon, A. *Wheel and Spindle*. Blackstaff Press, 1977. (Northern Ireland.)
Newman, Phil. *The Dartmoor Tin Industry: A Field Guide*. Chercombe Press, 1998.
Rogers, K. *Wiltshire and Somerset Woollen Mills*. Pasold, 1976.
Tann, Jennifer. *Gloucestershire Woollen Mills*. David & Charles, 1967.
Watts, Martin. *Corn Milling*. Shire, 1998.

Also the Shire albums on a variety of topics, including the following, contain much relevant information.

Aspin, Chris. *The Cotton Industry.* Shire, 1981; reprinted 1995.
Aspin, Chris. *The Woollen Industry.* Shire, 1982; reprinted 1994.
Atkinson, R. L. *Copper and Copper Mining.* Shire, 1987.
Baines, Patricia. *Flax and Linen.* Shire, 1985; reprinted 1998.
Bourne, Ursula. *Snuff.* Shire, 1990.
Crocker, Glenys. *The Gunpowder Industry.* Shire, second edition 1999.
Gale W. K. V. *Ironworking.* Shire, second edition 1998.
Willies, Lynn. *Lead and Leadmining.* Shire, 1982; reprinted 1999.

The Mills Section (formerly the Wind and Water Mill Section) of the Society for the Protection of Ancient Buildings (SPAB), London, has published a number of booklets, including:

Buckland, Stephen. *Lee's Patent Windmill 1744–1747.* SPAB, 1987.
Crocker, Glenys. *Gunpowder Mills Gazetteer.* SPAB, 1988.

The International Molinological Society (TIMS) publishes the Transactions of its symposia that have been held every four years since 1965. The papers cover a wide variety of subjects, all related to mills. TIMS also produces occasional publications under the title *Bibliotheca Molinologica.* See, for example:

Jones, David H. 'The Water-Powered Corn Mills of England, Wales and the Isle of Man', *Transactions of the Second Symposium,* Denmark (1969), 301–54.
Major, J. K. *The Windmills of John Wallis Titt.* TIMS, 1977.
Ward, Owen. *French Millstones.* TIMS, 1993.

Several of the regional and local mill groups produce interesting journals with a variety of articles concerned with research, recording and the repair of mills, and the Mills Research Group publishes the *Proceedings* of its annual conferences. Information on the mill groups is available from the Mills Section, SPAB, 37 Spital Square, London E1 6DY. Telephone: 020 7377 1644.

Publications concerned with the modern application of natural power, including practical handbooks, can be obtained from the Centre for Alternative Technology, Machynlleth, Powys SY20 9AZ, and a useful guide to contemporary wind farms is:

Hannah, Paul. *Wind Farms of the UK.* British Wind Energy Association, 1996.

PLACES TO VISIT

Much of the enjoyment of visiting historic and industrial sites can be to search out the remains of water- and wind-powered machinery that was used for a great variety of purposes. Some remains may be found as ruins, others converted to other uses, while many are preserved and maintained in working order. Properties owned or administered by English Heritage, Cadw: Welsh Historic Monuments, Historic Scotland, the National Trust and the National Trust for Scotland include many such sites. The Mills Section of the Society for the Protection of Ancient Buildings, 37 Spital Square, London E1 6DY (telephone: 020 7377 1644) publishes a regularly updated guide called *Mills Open* (sixth edition, 2000).

The following is a selection of sites that demonstrate the variety of uses to which water and wind power have been put. It is advisable to check opening arrangements in advance, particularly if travelling some distance.

General

Avoncroft Museum of Historic Buildings, Stoke Heath, Bromsgrove, Worcestershire B60 4JR. Telephone: 01527 831886. Website: www.avoncroft.org.uk (Danzey Green post mill.)
Morwellham Quay, Tavistock, Devon PL19 8JL. Telephone: 01822 832766. (Four waterwheels, including mine pumping, water pump and threshing.)
Museum of East Anglian Life, Stowmarket, Suffolk IP14 1DL. Telephone: 01449 612229. Website: www.suffolkcc.gov.uk/central/meal (Working corn mill and wind pump.)
Museum of Welsh Life, St Fagans, Cardiff CF5 6XB. Telephone: 029 2057 3500. Website: www.nmgw.ac.uk (Working corn mill, woollen mill, water-powered tannery, gorse mill and sawmill.)
Ulster Folk and Transport Museum, 153 Bangor Road, Cultra, Holywood, County Down BT18 0UE. Telephone: 028 9042 8428. Website: www.nidex.com/uftm (Spade mill, flax mill and corn mill.)
Weald and Downland Open Air Museum, Singleton, near Chichester, West Sussex PO18 0EU. Telephone: 01243 811348. Website: www.wealddown.co.uk (Working water-powered corn mill, wind pump, animal-powered machinery, treadwheel.)

Corn mills (water)

Alderholt Mill, Sandleheath Road, Fordingbridge, Hampshire SP6 1PU. Telephone: 01425 653130.
Bacheldre Mill, Church Stoke, Montgomery, Powys SY15 6TE. Telephone: 01588 620489. Website: www.go2.co.uk/bacheldremill
Barony Mills, Birsay, Orkney. Telephone: 01856 721439 or 771276.
Barry Mill, Barry, Carnoustie, Angus DD7 7RJ. Telephone: 01241 856761.
Blair Atholl Watermill, Blair Atholl, Pitlochry, Perthshire PH18 5SH. Telephone: 01796 481321.
Bromham Watermill, Bridgend, Bromham, Bedfordshire MK43 8LP. Telephone: 01234 824330.
Bunbury Mill, Mill Lane, off Bowesgate Lane, Bunbury, near Tarporley, Cheshire CW6 9PP. Telephone: 01829 261422.
Burcott Mill, Burcott, Wookey, Wells, Somerset BA5 1NJ. Telephone: 01749 673118.
Calbourne Watermill and Museum, Calbourne, Isle of Wight PO30 4JN. Telephone: 01983 531227.
Cann Mills, Cann, Shaftesbury, Dorset SP7 0BL. Telephone: 01747 852475.
Caudwell's Mill, Rowsley, Derbyshire. Telephone: 01629 73474.
Charlecote Mill, Hampton Lucy, Stratford-upon-Avon, Warwickshire CV35 8BB. Telephone: 01789 842072. Website: www.Geocities.com/Eureka/concourse/8261
Claybrooke Mill, Frolesworth Lane, Claybrooke Magna, Lutterworth, Leicestershire LE17 5DB. Telephone: 01455 202909.
Cotehele Mill, Cotehele, St Dominick, Saltash, Cornwall PL12 6TA. Telephone: 01579 351346.
Crabble Mill, Lower Road, River, Dover, Kent CT17 0UY. Telephone: 01304 823292.
Crakehall Mill, Crakehall, Bedale, North Yorkshire DL8 1HU. Telephone: 01677 423240.
Dalgarven Mill, Museum of Ayrshire Country Life and Costume, Dalry Road, Kilwinning, Ayrshire KA13 6PL. Telephone: 01294 552448.
Daniel's Mill, Eardington, Bridgnorth, Shropshire WV16 5JL. Telephone: 01746 762753.
Dunster Mill, Mill Lane, Dunster, Minehead, Somerset TA24 6SW. Telephone: 01643 821759.
Eling Tide Mill, Totton, Southampton, Hampshire SO40 9HF. Telephone: 01703 869575.
Eskdale Mill, Boot, Holmrook, Cumbria CA19 1TG. Telephone: 01946 723335.
Esk Mill, Danby, Whitby, North Yorkshire YO21 2JL. Telephone: 01287 660330.

Y Felin, St Dogmaels, Pembrokeshire SA43 3DY. Telephone: 01239 613999.
Ford End Watermill, Ford End Farm, Station Road, Ivinghoe, Buckinghamshire. Telephone: 01582 600391.
French Tidal Mill, Carew, near Tenby, Pembrokeshire SA70 8SL. Telephone: 01646 651782.
Gleaston Watermill, Gleaston, near Ulverston, Cumbria LA12 0QH. Telephone: 01229 869224. Website: www.watermill.co.uk
Headley Mill, Bordon, Hampshire GU35 8RJ. Telephone: 01420 472031.
Heatherslaw Mill, Ford, Cornhill-on-Tweed, Northumberland TD12 4TJ. Telephone: 01890 820338.
Heron Corn Mill, Beetham, Milnthorpe, Cumbria LA7 7AR. Telephone: 01 5395 65027.
Hinxton Watermill, Hinxton, Cambridgeshire. Telephone: 01223 243830.
Houghton Mill, Houghton, Huntingdon, Cambridgeshire PE17 2AZ. Telephone: 01480 301494. Website: www.nationaltrust.org.uk/houghtonmill
House Mill, Three Mill Lane, Bromley-by-Bow, London E3 3DU. Telephone: 0181 980 4626.
Kingsbury Watermill, St Michael's Street, St Albans, Hertfordshire AL3 4SJ. Telephone: 01727 853502.
Letheringsett Mill, Letheringsett, Holt, Norfolk NR25 7YO. Telephone: 01263 713153.
Little Salkeld Mill, Little Salkeld, Penrith, Cumbria CA11 1NN. Telephone: 01768 881523.
Lode Mill, Anglesey Abbey, Lode, Cambridgeshire CB5 9EJ. Telephone: 01223 811200.
Mapledurham Mill, Mapledurham, Reading, Berkshire RG4 7TR. Telephone: 0118 972 3350.
Maxey Mill, Mill Road, Maxey, Peterborough, Cambridgeshire PE6 9EZ. Telephone: 01778 343191.
Mill Green Museum and Mill, Mill Green, Hatfield, Hertfordshire AL9 5PD. Telephone: 01707 271362.
Muncaster Mill, Ravenglass, Cumbria CA18 1ST. Telephone: 01229 717232.
Museum of East Anglian Life, Stowmarket, Suffolk IP14 1DL. Telephone: 01449 612229. Website: www.suffolkcc.gov.uk/central/meal (Alton Mill.)
Museum of Welsh Life, St Fagans, Cardiff CF5 6XB. Telephone: 029 2057 3500. Website: www.nmgw.ac.uk (Melin Bompren.)
Nether Alderley Mill, Nether Alderley, near Alderley Edge, Macclesfield, Cheshire. Telephone: 01625 523012.
Old Mill, Mill Lane, Lower Slaughter, Gloucestershire. Telephone: 01451 820052.
Otterton Mill, Otterton, Budleigh Salterton, Devon EX9 7HG. Telephone: 01395 568521.
Pakenham Mill, Grimstone End, Pakenham, Suffolk IP31 2LR. Telephone: 01787 247179.
Quetivel Mill, St Peters Valley, St Peter, Jersey. Telephone: 01534 745408.
Redbournbury Mill, Redbourn, St Albans, Hertfordshire AL3 6RS. Telephone: 01582 792874.
Sarehole Mill, Colebank Road, Hall Green, Birmingham B13 0BO. Telephone: 0121 777 6612.
Shugborough Estate, Milford, near Stafford ST17 0XB. Telephone: 01889 881338. Website: www.staffordshire.gov.uk
Stainsby Mill, Hardwick Hall, Doe Lea, Chesterfield, Derbyshire S44 5QJ. Telephone: 01246 850430.
Stretton Mill, near Farndon, Chester, Cheshire. Telephone: 01606 41331.
Tockett's Mill, Skelton Road, Guisborough, North Yorkshire TS14 6QA.
Weald and Downland Open Air Museum, Singleton, near Chichester, West Sussex PO18 0EU. Telephone: 01243 811348. Website: www.wealddown.co.uk (Lurgashall Watermill.)
Wellesbourne Mill, Kineton Road, Wellesbourne, Warwick CV35 9HG. Telephone: 01789 470237.
Worsbrough Mill Museum, Worsbrough, Barnsley, South Yorkshire. Telephone: 01226 774527.

Corn mills (wind)
(P) = post mill; (S) = smock mill; (T) = tower mill.
Alford – The Five-Sailed Windmill (T), Alford, Lincolnshire LN13 9EQ. Telephone: 01507 462136.
Ashton Mill (T), Chapel Allerton, Wedmore, Somerset. Telephone: 01278 435399.
Avoncroft Museum of Historic Buildings, Stoke Heath, Bromsgrove, Worcestershire B60 4JR. Telephone: 01527 831886. Website: www.avoncroft.org.uk (Danzey Green post mill.)
Aythorpe Roding Mill (P), near Leaden Roding, Essex. Telephone: 01621 828162.
Ballycopeland Mill (T), Millisle, Newtownards, County Down. Telephone: 028 9054 3037.
Bembridge Mill (T), Bembridge, Isle of Wight PO35 5SQ. Telephone: 01983 873945.
Bourn Mill (P), Bourn, Cambridge. Telephone: 01223 243830.
Bursledon Mill (T), Windmill Lane, Bursledon, Hampshire SO31 8BG. Telephone: 02380 404999.
Buttrum's Mill (T), Woodbridge, Suffolk IP12 4JJ. Telephone: 01473 583352.
Cromer Mill (P), Cromer, near Stevenage, Hertfordshire SG2 7AD. Telephone: 01279 843301.
Downfield Mill (T), Soham, Ely, Cambridgeshire. Telephone: 01353 720333.
Draper's Mill (S), St Peter's Footpath, off College Road, Margate, Kent. Telephone: 01843 226227.
Ellis's Mill (T), Mill Road, Lincoln. Telephone: 01522 523870 or 528448.

Great Bircham Mill (T), Great Bircham, King's Lynn, Norfolk PE31 6SJ. Telephone: 01485 578393.
Great Gransden Mill (P), Great Gransden, Cambridgeshire SG19 3AA. Telephone: 01767 677283.
Green's Mill (T), Windmill Lane, Sneinton, Nottingham NG2 4QB. Telephone: 0115 915 6878.
Heckington Mill (T), Heckington, Sleaford, Lincolnshire. Telephone: 01529 461919.
High Salvington Mill (P), Worthing, West Sussex. Telephone: 01903 260218.
Jill Mill (P), Clayton, West Sussex. Telephone: 01273 843263.
Kibworth Harcourt Mill (P), Kibworth Harcourt, Market Harborough, Leicestershire. Telephone: 0116 279 2413.
Lacey Green Windmill (S), Lacey Green, near Princes Risborough, Buckinghamshire. Telephone: 01844 343560. Website: www.chilternsociety.freeserve.co.uk
Marsh Mill (T), Thornton Cleveleys, Blackpool, Lancashire. Telephone: 01253 860765.
Maud Foster Mill (T), Boston, Lincolnshire. Telephone: 01205 352188.
Melin Llynon (T), Llanddeusant, Anglesey. Telephone: 01407 730797.
Mountnessing Mill (P), Mountnessing, near Brentwood, Essex. Telephone enquiries: 01621 828162.
Mount Pleasant Windmill (T), Kirton-in-Lindsey, Lincolnshire DN21 4NH. Telephone: 01652 640177.
North Leverton Mill (T), Mill Lane, North Leverton, Retford, Nottinghamshire DN22 0AB. Telephone: 01427 880573.
Nutley Mill (P), Crowborough Road, Nutley, Uckfield, East Sussex. Telephone: 01435 873367.
Outwood Mill (P), Outwood Common, near Redhill, Surrey RH1 5PW. Telephone: 01342 843644. Website: www.cix.co.uk/injimnutt
Over Mill (T), Longstanton Road, Over, Cambridgeshire CB4 5PP. Telephone: 01954 230742.
Pakenham Mill (T), Pakenham, Suffolk. Telephone: 01359 230275.
Pitstone Windmill (P), Pitstone, Leighton Buzzard, Bedfordshire. Telephone: 01582 872303.
Polegate Mill and Museum (T), The Croft, Polegate, East Sussex. Telephone: 01323 731514 or 734496.
Quainton Mill (T), Quainton, Aylesbury, Buckinghamshire. Telephone: 01296 655348.
Sarre Mill (S), Canterbury Road, Sarre, Kent CT7 0JU. Telephone: 01843 847573.
Saxtead Green Mill (P), Saxtead Green, Framlingham, Woodbridge, Suffolk IP13 9QQ. Telephone: 01728 685789.
Shipley Windmill (S), Shipley, West Sussex RH13 8PL. Telephone: 01403 730439.
Skidby Windmill (T), Cottingham, East Riding of Yorkshire HU16 5TE. Telephone: 01482 884971.
Stansted Mountfitchet Mill (T), Millside, Stansted Mountfitchet, Essex. Telephone: 01279 813160.
Stanton Mill (P), Stanton, Bury St Edmunds, Suffolk. Telephone: 01359 250622.
Stelling Minnis Mill (S), Canterbury, Kent CT4 6AQ. Telephone: 01227 709550 or 709419.
Stembridge Mill (T), High Ham, Langport, Somerset TA10 9DL. Telephone: 01458 250818.
Sutton Mill Broads Museum (T), Stalham, Norfolk NR12 9RZ. Telephone: 01692 581195.
Swaffham Prior Mill (T), Swaffham Prior, Cambridgeshire CB5 0JZ. Telephone: 01638 741009.
Thaxted Mill (T), Thaxted, Dunmow, Essex. Telephone: 01371 830285.
Thelnetham Mill (T), Mill Road, Thelnetham, Suffolk IP22 1JZ. Telephone: 01359 250622.
Trader Mill (T), Sibsey, Boston, Lincolnshire PE22 0SY. Telephone: 01205 820065.
Tuxford Mill (T), Harold Cottage, East Markham, Newark, Nottinghamshire NG22 0QM. Telephone: 01777 870413.
Union Mill (S), The Hill, Cranbrook, Kent. Telephone: 01580 712984.
Upminster Mill (S), Upminster, Essex. Telephone: 01708 447535.
West Blatchington Mill (S), Hove, East Sussex. Telephone: 01273 776017.
Whissendine Mill (T), Melton Road, Whissendine, Rutland LE15 7EU. Telephone: 01664 474172.
Wilton Windmill (T), Wilton, Marlborough, Wiltshire SN8 3SS. Telephone: 01672 870266.
Wimbledon Windmill and Museum (S), Wimbledon Common, London SW19 5NR. Telephone: 020 8947 2825. Website: www.wpcc.org.uk
Wrawby Mill (P), Wrawby, Brigg, Lincolnshire. Telephone: 01652 653699.

Drainage and pumping windmills

Berney Arms Windmill, Great Yarmouth, Norfolk. Telephone: 01493 700605.
Boardman's and Clayrack Drainage Mills, How Hill, Ludham, Norfolk. Telephone: 01603 222705.
Elvington Brickyard Wind Pump, Elvington Lake Cottage, Elvington, North Yorkshire. Telephone: 01904 608255.
Herringfleet Drainage Mill, Lowestoft, Suffolk. Telephone: 01473 583352.
Museum of East Anglian Life, Stowmarket, Suffolk IP14 1DL. Telephone: 01449 612229. Website: www.suffolkcc.gov.uk/central/meal (Eastbridge Wind Pump.)
Stracey Arms Drainage Pump, Acle, Norfolk. Telephone: 01603 222705.
Thurne Dyke Wind Pump, Norfolk. Telephone: 01603 222705.

Weald and Downland Open Air Museum, Singleton, near Chichester, West Sussex PO18 0EU. Telephone: 01243 811348. Website: www.wealddown.co.uk (Westham Wind Pump.)
Wicken Fen Pumping Mill, Wicken, Ely, Cambridgeshire CB7 5XP. Telephone: 01353 720274. Website: www.wicken.org.uk

Water-powered pumps
Castleton Water Pumping Station, Oborne Road, Sherborne, Dorset. Telephone: 01935 813391.
Claverton Pumping Station, Claverton, Bath, Somerset. Telephone: 01225 483001.
Coultershaw Beam Pump, Petworth, West Sussex. Telephone: 01798 865569.
Painshill Park, Portsmouth Road, Cobham, Surrey KT11 1JE. Telephone: 01932 868113.

Woollen mills
Bradford Industrial Museum, Moorside Road, Eccleshill, Bradford, West Yorkshire BD2 3HP. Telephone: 01274 631756.
Brynkir Woollen Mill, Golan, near Porthmadog, Gwynedd LL51 9YU. Telephone: 01766 530236.
Coldharbour Mill, Uffculme, Devon EX15 3EE. Telephone: 01884 840960.
Higher Mill, Holcombe Road, Helmshore, Rossendale, Lancashire BB4 4NP. Telephone: 01706 226459.
Leeds Industrial Museum, Armley Mill, Canal Road, Armley, Leeds, West Yorkshire LS12 2QF. Telephone: 0113 263 7861. Website: www.leeds.gov.uk
Melin Tregwynt Woollen Mill, Castle Morris, Haverfordwest, Pembrokeshire SA62 5UX. Telephone: 01348 891225.
Museum of the Welsh Woollen Industry, Drefach Felindre, Llandysul, Carmarthenshire SA44 5UP. Telephone: 01559 370929.
Penmachno Woollen Mill, Betws-y-Coed, Conwy. Telephone: 01690 710545.
Rock Mill, Capel Dewi, Llandysul, Ceredigion SA44 4PH. Telephone: 01559 362356.
Trefriw Woollen Mill, near Betws-y-Coed, Conwy LL27 0NQ. Telephone: 01492 640462.

Cotton mills
Sir Richard Arkwright's Cromford Mill, Mill Lane, Cromford, Derbyshire DE4 3RQ. Telephone: 01629 823256.
New Lanark Mills, Lanark, Lanarkshire ML11 9DB. Telephone: 01555 661345.
Quarry Bank Mill, Styal, Cheshire SK9 4LA. Telephone: 01625 527468.

Silk mill
Whitchurch Silk Mill, Whitchurch, Hampshire RG28 7AL. Telephone: 01256 892065.

Linen mills
Ulster Folk and Transport Museum, 153 Bangor Road, Cultra, Holywood, County Down BT18 0EU. Telephone: 028 9042 8428. Website: www.nidex.com/uftm (Gorticashel Flax Mill.)
Wellbrook Beetling Mill, Corkhill, Cookstown, County Tyrone. Telephone: 028 8675 1735.

Mining and extractive industries
Dolaucothi Gold Mines, Pumsaint, Llanwrda, Carmarthenshire SA19 8RR. Telephone: 01558 650359.
Killhope Lead Mining Centre, near Cowshill, Upper Weardale, County Durham DL13 1AR. Telephone: 0191 383 3337.
Llywernog Silver Lead Mine, Ponterwyd, Aberystwyth, Cardiganshire. Telephone: 01970 890620.
Lady Isabella Wheel, Laxey, Isle of Man. Telephone: 01624 648000.
Minera Lead Mine, Wern Road, Minera, Wrexham LL14 4HT. Telephone: 01978 261529.
Morwellham Quay, Tavistock, Devon PL19 8JL. Telephone: 01822 832766.
Tolgus Tin, New Portreath Road, Redruth, Cornwall TR16 4HN. Telephone: 01209 215185.
Welsh Slate Museum, Padarn Country Park, Llanberis, Gwynedd LL55 4TY. Telephone: 01286 870630.

China clay
Wheal Martyn Museum, Carthew, St Austell, Cornwall PL26 8XG. Telephone: 01726 850362.

Metalworking
Abbeydale Industrial Hamlet and Shepherd Wheel, Sheffield, South Yorkshire. Telephone: 0114 236 7731.
Anne of Cleves House and Museum, 52 Southover High Street, Lewes, East Sussex BN7 1JA. Telephone: 01273 474610.

Bersham Iron Works, Bersham Heritage Centre, Bersham, Wrexham LL14 4HT. Telephone: 01978 261529.
Bonawe Iron Furnace, Taynuilt, Argyll PA65 1JE. Telephone: 01866 822432.
Churchill Forge, Churchill, near Kidderminster, Worcestershire DY10 3LX. Telephone: 01562 700476.
Dyfi Furnace, Machynlleth, Powys SY20 8PG. Telephone: 029 2082 6185 for enquiries (monument is not staffed).
Finch Foundry, Sticklepath, Devon EX20 2NW. Telephone: 01837 840046.
Forge Mill Needle Museum, Needle Mill Lane, Riverside, Redditch, Worcestershire B98 8HY. Telephone: 01527 62509.
Patterson's Spade Mill, 751 Antrim Road, Templepatrick, County Antrim BT39 0AP. Telephone: 028 9443 3619.
Saltford Brass Mills, The Shallows, Saltford, Bristol. Telephone: 0117 986 2216.
Ulster Folk and Transport Museum, 153 Bangor Road, Cultra, Holywood, County Down BT18 0EU. Telephone: 028 9042 8428. Website: www.nidex.com/uftm (Coalisland Spade Mill.)
Wortley Top Forge, near Barnsley, South Yorkshire. Telephone: 0114 234 3479.

Flint and stone grinding
Cheddleton Flint Mills, Cheddleton, near Leek, Staffordshire ST13 7HL. Telephone: 01782 502907.
Mosty Lee Mill, Stone, Staffordshire. Telephone: 01785 619131.
Thwaite Mills, Great Lane, Stourton, Leeds, West Yorkshire LS10 1RP. Telephone: 0113 249 6453.

Saw and wood turning
Castle Ward Sawmill, Strangford, Downpatrick, County Down BT30 7LS. Telephone: 028 4488 1204.
Combe Mill, Woodstock, Oxfordshire. Telephone: 01993 811118.
Dunham Massey, Altrincham, Cheshire WA14 4SJ. Telephone: 0161 941 1025.
Florence Court, Enniskillen, County Fermanagh BT92 1DB. Telephone: 028 6634 8249.
Gunton Park Sawmill, Gunton Park, Norfolk. Telephone: 01603 222705.
Stott Park Bobbin Mill, Low Stott Park, Finsthwaite, Ulverston, Cumbria LA12 8AX. Telephone: 01539 531087.

Gunpowder
Chart Gunpowder Mills, Westbrook Walk, Faversham, Kent. Telephone for enquiries: 01795 534542.
Waltham Abbey Royal Gunpowder Mills, Waltham Abbey, Essex EN9 1BN. Telephone: 01992 767022. (Opening 2001.)

Paper
Wookey Hole Paper Mill, Wookey Hole, Wells, Somerset BA5 1BB. Telephone: 01749 672243. Website: www.wookey.co.uk

Modern water and wind power
Aberdulais Falls, near Neath SA10 8EU. Telephone: 01639 636674.
Centre for Alternative Technology, Machynlleth, Powys SY20 9AZ. Telephone: 01654 702400. Website: www.foe.co.uk/CAT
Cragside, Rothbury, Morpeth, Northumberland NE65 7PX. Telephone: 01669 620150.
Mary Tavy Power Station, Mary Tavy, Devon PL19 9PR. Telephone: 01822 810248.
National Dragonfly Museum (including Ashton Mill), Ashton, Peterborough PE8 5LE. Telephone: 01832 272264. (Turbines, hydroelectricity.)

Wind farms
Carno, near Newtown, Powys. Telephone: 01686 413973.
Oldside, Workington, Cumbria.
Penrhyddlan and Llydiartywaen, Dolfor, near Newtown, Powys.
Wind Electric Ltd, Delifarm, Delabole, Cornwall PL33 9BZ. Telephone: 01840 214100.

INDEX

Page numbers in italic refer to illustrations